Facundo Alejandro Sibona Bulfon
Guillermo Ricardo Catuogno
Juan Pablo Demichelis

Design and implementation of a PLC based on open technology

Facundo Alejandro Sibona Bulfon
Guillermo Ricardo Catuogno
Juan Pablo Demichelis

Design and implementation of a PLC based on open technology

Control and monitoring with Arduino, Node-RED and MQTT: affordable and scalable solution

ScienciaScripts

Imprint

Cover image: www.ingimage.com

This book is a translation from the original published under ISBN 978-613-9-43516-6.

Publisher:
Sciencia Scripts
is a trademark of
Dodo Books Indian Ocean Ltd. and OmniScriptum S.R.L publishing group

120 High Road, East Finchley, London, N2 9ED, United Kingdom
Str. Armeneasca 28/1, office 1, Chisinau MD-2012, Republic of Moldova, Europe
Printed at: see last page
ISBN: 978-620-8-29312-3

DEDICATION

This final project is dedicated to:

To my mother Andrea, who has taught me to trust myself, to be perseverant in the face of new goals and, above all, to be grateful and enjoy the processes in life.

To my father Alejandro, with whom I have the pleasure of sharing engineering projects, whose professionalism, responsibility and kindness represent my source of inspiration.

To my brothers Franco and Maia, for sharing so many moments together and for their incessant curiosity about the world around us, which generates an amazing learning circle.

To my grandparents Jacinta, Jorge and Ana, whose amazement at engineering motivates me to continue learning every day.

ACKNOWLEDGMENTS

I would like to express my deep gratitude to all the staff at the Universidad Nacional de San Luis for making this project possible and for providing a stimulating learning environment throughout my time there. From there, a special mention to my professors and thesis directors Juan Pablo and Guillermo, who with their dedication to their work and their impetus to transmit and teach, meant an extraordinary impulse during the career and the execution of this work. I would also like to remember and thank all the professors who were part of my journey in this institution.

My thanks to my co-workers, with whom I faced challenges and acquired the experience and wisdom that today define my professional career, and who through their support taught me the value of companionship.

To my friends in college and in life, who were by my side in critical moments and shared with me joys and achievements.

Last but not least, to all my family and the rest of the people who influenced my learning process, who with their words of encouragement, sincere advice and genuine joy have enlightened me and pushed me to pursue my dreams.

SUMMARY

This project deals with the design and implementation of a Programmable Logic Controller (PLC) based on open technology for the control and automation of a water treatment system. The design begins taking as a basis the Arduino development environment, which represents a tool for technological projects of all kinds widely spread around the world. From there, the design is oriented to convert this technological solution into a more robust one that adapts to industrial environments in an effective way. The paper has been divided into four sections. In Chapter 1, the proposal is detailed along with its scope and objectives, determined based on the automation requirements of the water treatment plant. It also describes the knowledge required to enter the world of programmable logic controllers and presents the open technology tools available as a solution. Chapter 2 describes the development of each design stage, from the selection of the microcontroller and the electronic components, to the design of the printed circuit board and the 3D modeling of the housings. Ultimately, the control logic and programming code used is discussed in detail. The cost analysis in Chapter 3 covers a comprehensive evaluation of the costs associated with implementing the device in the project. Costs related to the procurement of electronic components, the fabrication of the 3D housings and the programming of the system are included in order to obtain an accurate picture of the economic investment involved. Finally, Chapter 4 presents the conclusions, synthesizing the findings and results obtained during the development of this mechatronic project. The fulfillment of the objectives set is evaluated, and the limitations encountered are analyzed along with possible areas of improvement for future developments.

Keywords - Arduino, automation, programmable logic controller, electronics, 3D modeling, programming, open technology, communication protocol.

TABLE OF CONTENTS

CAPITULO 1: PROPOSAL

1.1 Introduction

Industrial automation has become a fundamental pillar for the efficiency and optimization of production processes and to facilitate tasks in various areas, from production lines in the manufacturing industry to control systems in intelligent buildings. In this context, Programmable Logic Controllers (PLC) have been fundamental pieces, being in charge of interpreting the environment and acting accordingly in an automatic and accurate way.

Despite their remarkable success and distribution, traditional PLCs have certain limitations in certain applications. They are expensive and require specialized knowledge for programming, which makes them inaccessible for small companies or independent projects. This is where Arduino technology comes in, which, with its open source approach and active community, provides a more affordable and adaptable alternative for automation. The combination of the robustness of PLCs and the flexibility of Arduino opens up new opportunities for the creation of customized and scalable control systems.

The trend towards the use of open software and open hardware is gaining popularity, as it offers numerous advantages in terms of flexibility, customization and cost reduction. On the other hand, industrial protocols ensure efficient and reliable communication between field devices and control systems, enabling real-time data collection and precise control of industrial processes. This combination creates a robust and efficient infrastructure for automation and monitoring.

This convergence between open technologies and industrial communication protocols is driving digital transformation in industry, fostering innovation and improving operational efficiency. In this context, the development of solutions based on open PLCs presents itself as a promising option for companies seeking to remain competitive in a constantly evolving industrial environment.

1.2 Objectives

1.2.1 General objectives is

- Develop and implement a low-cost Programmable Logic Controller (PLC) using open technologies.
- To use the device in an industrial environment for the effective control of a water treatment equipment, with a permanent operating regime.
- Interconnect the device with other systems through the implementation of industrial protocols, for information gathering and to facilitate the creation of automation networks.

1.2.2 Specific objectives

- Design the PLC electronic circuit, capable of operating in various industrial environments.
- Select components and design the printed circuit board (PCB) for the controller manufacturing, considering cost and efficiency aspects.
- Implement digital and analog input modules, as well as relay and optocoupled output modules, to cover various control needs.
- 3D modeling of the PLC housing, ensuring an ergonomic and functional design.
- Integrate I2C and UART communication protocols to enable efficient interfacing with other devices and systems.
- Use a microcontroller from the range of options offered by Arduino and similar, optimizing performance at low cost and taking advantage of its wide acceptance in the community.
- Implement an industrial communication module to transmit data from the sensors and the PLC to a server, which will allow remote monitoring and control of the process from any device with an internet connection.

1.3 Scope and limitations

The project covers from the initial conception and design of the PLC to the manufacturing of the functional prototype, including the development of the electronic circuit, component selection, PCB design, implementation of input and output modules, 3D modeling of the housing and integration of communication protocols. The focus is, on the one hand, on the economic viability and democratization of industrial automation, making the developed technology accessible and transparent, and on the other hand, on the connectivity with different systems for remote data processing and monitoring, from any device with internet connection.

Printed circuit board robustness will be a challenge, and although optimization will be sought, absolute resistance to all industrial environments cannot be guaranteed. Functional testing will be limited to visual analysis of device performance in the industrial environment. The project focuses on the use of microcontrollers within the range of Arduino and similar options, limiting the possible specific features of other more specialized microcontrollers. Integration with specific industrial automation systems may require additional adaptations that are not contemplated in this project.

1.4 Theoretical framework

1.4.1 Programmable Logic Controllers (PLC)

1.4.1.1 Definition of PLCs

A programmable logic controller, better known by its acronym PLC (Programmable Logic Controller), is a computer-like device used in industrial automation to automate electromechanical processes, such as the control of factory machinery on assembly lines or mechanical attractions. NEMA (National Electrical Manufacturers Association) defines PLC as:

"Electronic instrument, which uses programmable memory to store instructions on the implementation of certain functions, such as logical operations, sequences of actions, time specifications, counters and calculations for control by means of analog or digital I/O modules on different types of machines and processes."

A digital electronic device specifically designed to be easily programmed. This type of processor is called a logic processor because programming is primarily concerned with the execution of logic and switching operations. The input devices (sensors and switches) and output devices (actuators) that are under control are connected to the PLC, and then the controller monitors the inputs and outputs according to the program stored by the operator in the PLC with which it controls machines or processes (Figure 1).Figure N° 1 1). These have the great advantage that they allow you to modify a control system without having to rewire the connections of input and output devices; just the operator to type on a keyboard the corresponding instructions. The result is a flexible system that can be used to control systems that are very diverse in nature and complexity. Such systems are widely used for the implementation of control logic functions because they are easy to use and program [1].

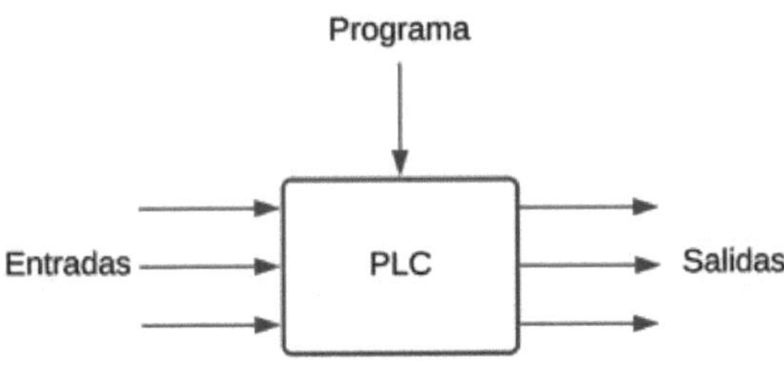

Figure N° 1 1Schematic diagram of a PLC

PLCs are similar to computers, but have specific features that allow them to be used as controllers. These characteristics are:

1. They are robust and designed to withstand vibration, temperature, humidity and noise.
2. The interface for inputs and outputs is inside the controller.
3. It is very easy to program them.

Figure N° 2 2Siemens PLC Logo

In the Figure N° 2 2 the typical format of a PLC can be observed, which corresponds to the Siemens LOGO model. Although it is a model with basic functions, it has the essential characteristics that a PLC must have. We can highlight in the upper part the inputs for the power supply of the device and the signal inputs, while in the lower part we can see the connections for the outputs. On the front is the Ethernet port to connect to the programmer, through which the desired program is loaded, and in some, you can find screens and buttons to interact with it.

1.4.1.2 Basic PLC structure

The Figure N° 3 3 shows the basic internal structure of a PLC which, in essence, consists of a central processing unit (CPU), memory and input/output circuits. The CPU controls and processes all operations within the PLC. It has a timer whose typical frequency is between 1 and 8 MHz. This frequency determines the operating speed of the PLC and is the source of timing and synchronization of all elements of the system. A bus system carries information and data to and from the CPU, memory and input/output units. The memory elements are: a ROM for permanent storage of operating system information and corrected data; a RAM for the user program and temporary buffer memory for the input/output channels.

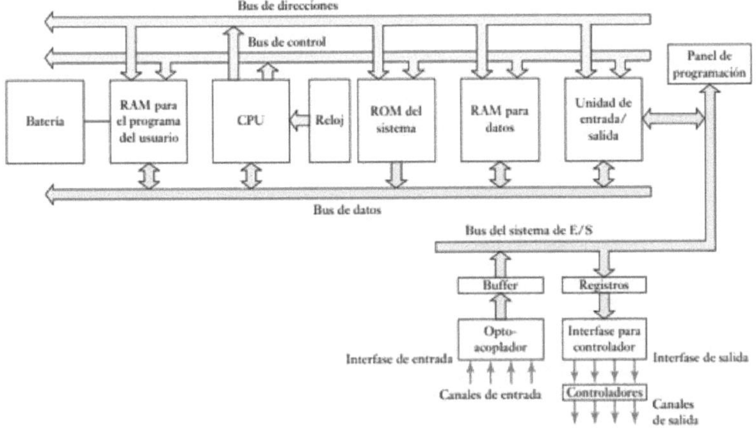

Figure N° 3 3PLC architecture

1.4.1.3 Inputs/Outputs

The input/output unit is the interface between the system and the external world and where the processor receives information from external devices and communicates information to external devices. Input/output interfaces provide isolation and signal conditioning functions so that those sensors and actuators can often be connected directly to them without requiring other circuitry. Inputs can range from switches that are triggered by an event, or other sensors such as temperature sensors or flow sensors. Outputs can be used to activate motor starter coils, solenoid valves, etc. Electrical isolation from the external world is usually by means of optoisolators. The Figure N° 4 4 shows the basic form of an input channel. The digital signal that is usually compatible with the microprocessor in the PLC is 5 VDC. However, signal conditioning on the input channel, with isolation, increases the range of input signals. Therefore, PLCs with input voltages of 5 V, 24 V, 110 V and 240 V can be found.

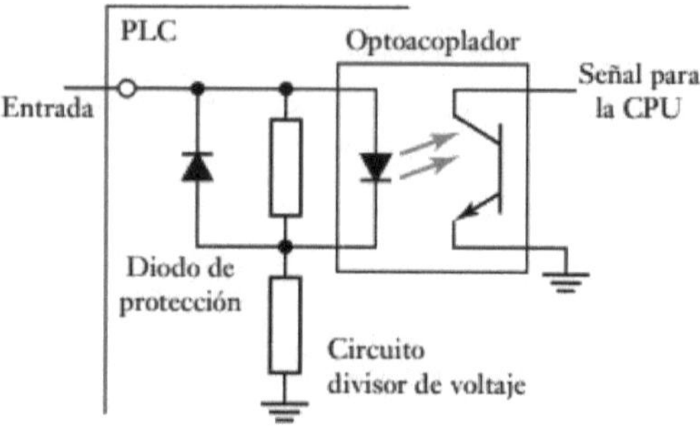

Figure N° 4 4Input channel

Outputs can be of the type:

- Relay or relay
- Transistor
- Triac

With the relay type, the PLC output signal is used internally to operate a relay to switch currents within a few amperes in an external circuit. The relay isolates the PLC from the external circuit and can be used to switch direct current (DC) or alternating current (AC). However, relays are relatively slow to operate.

The output transistor type uses a transistor to switch the current in the external circuit. This causes a faster switching action. Optoisolators are used with transistor switches to cause isolation between the external circuit and the PLC. The transistor output is for DC switching only.

Triac type outputs can be used to control external AC loads. Optoisolators are again used to provide isolation, allowing the outputs to be a 24 V and 100 mA signal, a 110 V and 1 A DC voltage, or a 240 V and 1 A or 2 A AC voltage.

1.4.1.4 Input programs

Programs are entered into the input/output unit from small hand-programmed devices, desktop consoles with a display unit, keyboard and screen display or via a link to a personal computer (PC) that is loaded with an appropriate software package. Only when the program has been designed on the programming device and is ready, it is transferred to the PLC memory unit. Once a program has been developed in RAM it can be loaded into an EPROM chip and made permanent. Specifications for small PLCs often detail the size of the program memory in terms of program steps that can be stored. A program step is an instruction for some event to occur. A program task may consist of a number of steps and may be, for example: examine the state of switch A, examine the state of switch B, if A and B are closed, then energize solenoid valve P which may result in the operation of some actuator. When this happens another task is initiated. It is common for the number of steps a small PLC can handle to be 300 to 1000, which is generally adequate for most control situations.

1.4.1.5 PLC hardware components

A PLC may contain a cassette with a track containing various types of modules, as shown in the following figure Figure N° 5corresponding to a Siemens PLC:

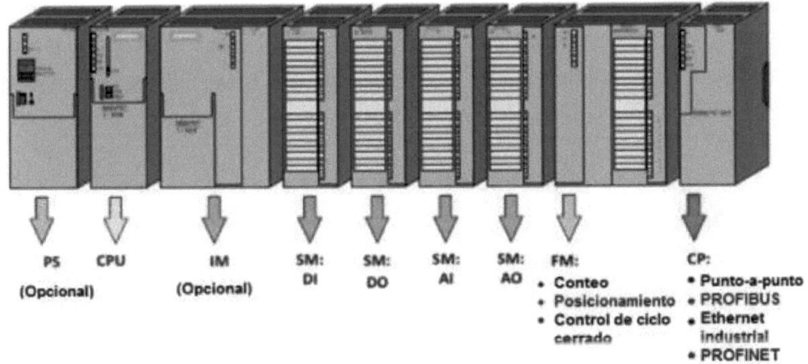

Figure N° 5PLC hardware components

PS (Power Supply): This is the power supply, which converts electrical energy from the external power source (such as the mains) into the form of voltage and current needed to power the internal components of the PLC .

CPU: It is the Central Processing Unit, which is the PLC itself.

IM: Interface Module, which connects different modules to the same PLC.

SM (Standard Module): These are the standard input/output modules. They can be:

- DI: Digital Inputs

- DO: Digital Outputs
- AI: Analog Input
- AO: Analog Output

FM (Functional Module): FM modules are typically specialized modules that perform specific functions or provide additional features beyond the standard input/output capabilities. For example, fast counting, PID controller or position control.

CP (Communication Processor): It is the module in charge of connecting the PLC to an industrial network under different protocols, such as Ethernet, PROFIBUS, point-to-point serial connection, etc.

HMI (Human Machine Interface): It is the device in charge of providing ways to interact between man and PLC, such as screens with keyboard. Each PLC module has its own basic HMI interface, used for displaying errors and communication conditions, battery, inputs/outputs, PLC operation, etc. Small liquid crystal displays (LCD) or light emitting diodes (LED) are used for the HMI interface.

1.4.1.6 Types of signals used by PLCs

A PLC receives and transfers electrical signals representing finite physical variables (temperature, pressure etc.). Thus it is necessary to include in the SM a signal converter to receive and change the values to physical variables. There are three types of signals in a PLC: binary, digital and analog signals. [2].

1) <u>Binary signals</u>: one-bit signal with two possible values ("0" - low level, false or "1" - high level, true), which are encoded by means of a button or a switch. An activation normally opens the contact corresponding to the logic value "1", and a non-activation to the logic level "0". Thus IEC 61131 defines the range -3 - +5 V for the logic value "0", while 11 - 30 V are defined as the logic value of "1" (for non-contact sensors) at 24 V DC. In addition, at 230 V AC, IEC 61131 defines the range 0 - 40 V for the logic value of "0", and 164 - 253 V for the logic value "1".

2) <u>Digital signals</u>: this is a sequence of binary signals, considered as one. Each position of the digital signal is called a bit. Typical formats of digital signals are: tetrad - 4 bits (rarely used), byte - 8 bits, word - 16 bits, double word - 32 bits, double long word - 64 bits (rarely used).

3) Analog signals are signals that have continuous values, i.e. consist of an infinite number of values (e.g. in the range 0 - 10 V). Nowadays, PLCs cannot process real analog signals. Thus, these signals must be converted into digital signals and vice versa. This conversion is done by means of analog measuring sensors (SM), which contain analog-to-digital converters (ADC). The high resolution and accuracy of the analog signal can be achieved by using more bits in the digital signal. For example, a typical analog signal of 0 - 10 V can be converted with an accuracy of 0.1 V, 0.01 V or 0.001 V, depending on the number of bits in the digital signal. Another popular industrial analog signal is the 4-20mA current loop. This is a type of analog signal in which the magnitude of the electrical current varies proportionally with the process variable being measured or controlled. In this range, 4 mA usually corresponds to the minimum value of the process, while 20 mA usually corresponds to the maximum value. It has the advantage of being less susceptible to electromagnetic interference and electrical noise compared to voltage signals, and can be transmitted over long distances without significant signal degradation.

1.4.1.7 Principle of operation

A PLC operates cyclically, as described below:

1) Each cycle starts with internal PLC maintenance work such as memory control, diagnostics, etc. This part of the cycle is executed very quickly so that the user does not notice it.

2) The next step is the update of the inputs. The input conditions of the SMs are read and converted into binary or digital signals. These signals are sent to the CPU and stored in memory data.

3) The CPU then executes the user program, which has been sequentially loaded into memory (each instruction individually). New output signals are generated during program execution.

4) The last step is the updating of the outputs. After execution of the last part of the program, the output signals (binary, digital or analog) are sent to the SM from the data in memory. These signals are then converted into the appropriate signals for the actuator signals. At the end of each cycle the PLC starts a new cycle.

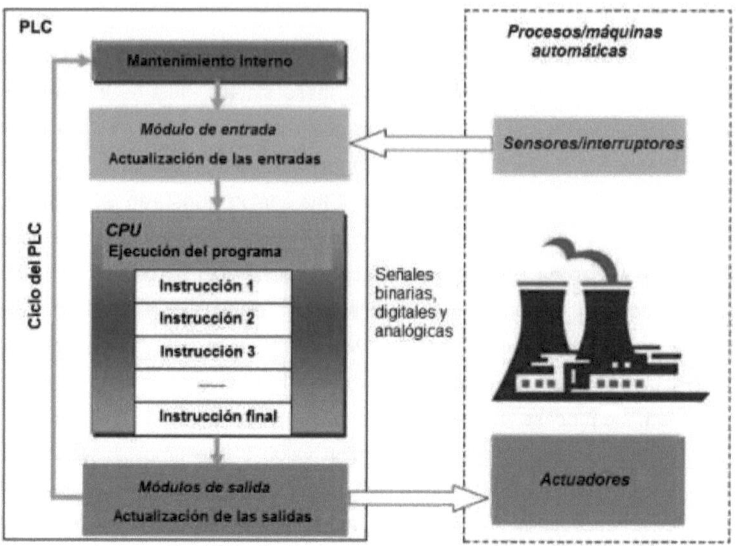

Figure N° 6 6Siemens S7-300 PLC Operation Cycle

1.4.2 Industrial communications

Industrial communications is broadly defined as a data transmission system designed to enable communication between field devices such as sensors, actuators, controllers and other equipment and monitoring and control systems such as process control systems or management systems.

The signals between periphery and control, initially analog and point-to-point, thanks to the development of digital electronics and the rise of microprocessors, became a set of signals capable of carrying information through a single transmission medium (Fieldbus).

Among the advantages that the use of this communication offers are the possibility of remote programming, remote supervision, diagnostics of all connected elements, modularity, access to information practically instantaneously, etc.

The IEC 61158 fieldbus standard gives the following definition: "A fieldbus is a digital, serial, multidrop data bus for communication with industrial control devices and instrumentation such as - but not limited to - transducers, actuators and local controllers." [3]

1.4.2.1 Communication levels in an industrial network

The integration of the different equipment and devices in an industry is done by dividing the tasks into groups of processors with a hierarchical organization. Thus, depending on the function and type of connections, five levels are usually distinguished in an industrial network: [4]

N1 - Input/output level: it is the level closest to the process. This level is basically made up of signal reception, actuation and data input/output units of the process or of a local operator. Normally these types of networks are characterized by short message lengths, message reliability and integrity, the efficiency of the protocol used, transmission speed and the medium access technique of the devices.

N2 - Field level: integrates small automation devices (compact PLCs, PIDs, I/O multiplexers, etc.) into sub-networks or 'islands'. At the highest level of these networks, one or more modular PLCs can be found acting as network masters. At these first two levels, so-called fieldbuses are used, which are the simplest and most process-oriented level within the structure of industrial communications.

N3 - Process control level: This level consists of control units (with their own CPU and programs) such as automatons, process controllers, robot controllers, numerical controllers, etc., which are responsible for the automatic control of certain parts of the plant. The network integration of these units allows the exchange of data and information useful for the global control of the process. It is at this level that LAN type networks (MAP or Ethernet) are usually used.

N4 - Production control level: This level includes a series of units intended for the overall control of the process, such as process computers, dialog terminals, terminals for liaison with other departments of the company, etc. From these units there is access to most of the process variables, generally for the purpose of supervising them, presenting them, recording and/or storing them, changing set points, altering programs and obtaining data for further processing.

N5 - Management level: This level includes communication with management computers and is responsible for the processing of data, obtained at the previous level, and its use in statistical analysis, manufacturing control, quality control, inventory management and general management. In some cases, units at this level may have connections to larger proprietary WAN-type networks and/or Internet broadcast standards.

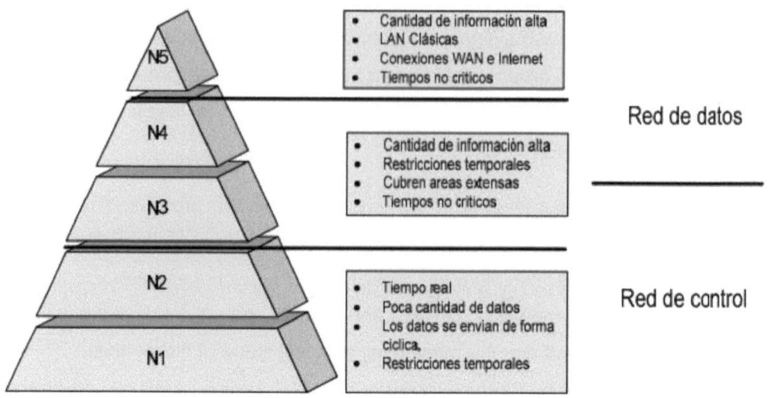

Figure N° 7 7Hierarchy of industrial communications

1.4.2.2 Data transmission modes

1.4.2.2.1 Parallel data transmission

Parallel data buses transmit 8, 16 or 32 bits simultaneously. In this type of communication, each data bit and each control signal has a dedicated bus line. To transmit a complete 8-bit word, 8 data lines are required. Along with these data lines, flow control lines are also required. Parallel data transmission allows high data transfer speeds, but its cabling and interfaces are more expensive than those of serial transmission. The trade-off is that this type of connection is very vulnerable to electromagnetic interference and is therefore used for very short data transmission distances. [4]

It is measured in bits, or lines of communication. Thus we have buses of 8, 16, 32, 64, 128 bits.

1.4.2.2.2 Serial data transmission

Data transmission is carried out bit by bit, sequentially, on the same line, together with the transmission control bits, using only two conductors. Since the blocks of information are transmitted sequentially and not simultaneously, the data transfer rate is much lower than in the case of parallel data transmission for the same bit transfer rate. However, this type of transmission is more economical. [4]

Two methods can be used to synchronize transmitter and receiver:

Asynchronous: Sender and receiver work at the same speed and with the same number of bits per message. A certain signal (start bit) indicates the start of the message, and the receiver starts sampling the signal present on the medium. This method requires precision in the sampling operations (constant clock periods over time).

Synchronous: An additional clock signal indicates the signal sampling times to the receiver. This method requires an additional communication line. The advantage of this method is that the receiver only has to follow the edges of the clock signal.

1.4.3 Communication protocols

Once we have defined the physical support and the characteristics of the signal to be transmitted, we must determine the way in which the information exchange will be carried out (synchronization between the line ends, error detection and correction, communication link management, etc.).

The communication protocol encompasses all the rules and conventions that any two computers must follow in order to exchange information. [5]

1.4.4 Communication Protocols in Industrial Automation

Overview of the most commonly used protocols in control and automation systems.

1.4.4.1 Serial protocols

RS-232: Designed for communications between a control terminal and a system in the field, such as a sensor. It is limited to short distances and low transmission speeds, connecting only two devices (ideal for point-to-point networks). The connectors used are limited to DB-25 and DB-9.

Figure N° 8 8DB-9 connector for RS-232 communications

RS422: increases the range (transmission distance) and speed with respect to the RS232 protocol, as well as being more resistant to electromagnetic noise, but its main virtue is that it can be used in point-to-point or multipoint topologies, such as star networks. It is not limited in terms of the connectors that can be used.

RS485: it is also multidrop and can operate over long distances. As it has two wires unlike RS422, it can connect multiple controllers with multiple devices over the field, although this leads to its programming not being as simple as the previous two.

1.4.4.2 Fieldbus protocols

The Foundation Fieldbus Association provides the following definition for fieldbuses:

"A digital, bi-directional, multi-access communication link that enables communication between smart meters and control devices. This is served as a local area network for advanced control processes, remote inputs and outputs, and high-speed automation applications." [5]

Fieldbuses replace traditional point-to-point wiring with a single communication line, simplifying design and reducing system cost. They enable real-time transmission of data and commands, which improves the efficiency and reliability of industrial control systems.

The following are the most commonly used protocols in industrial networks:

1.4.4.2.1 *Profibus (Field Bus Process)*

It is a fieldbus standard, developed in 1987 by the German companies Bosch, Klöckner Möller and Siemens. It is an open, standard, manufacturer-independent, multi-profile network that adapts to the conditions of industrial automation applications. [5]

The profiles are as follows:

- Profibus FMS (Fieldbus Message Specification): It is oriented to the exchange of large amounts of data between PLCs. In this type of transmission, functionality is more important than speed, so reaction times are slower. Generally, data transmission is acyclic (program-controlled). [6].
- Profibus DP (Decentralized Periphery): A high-speed solution designed and optimized for communication between automation systems and distributed devices.
- Profibus PA (Process Automation): It is a process application oriented solution.

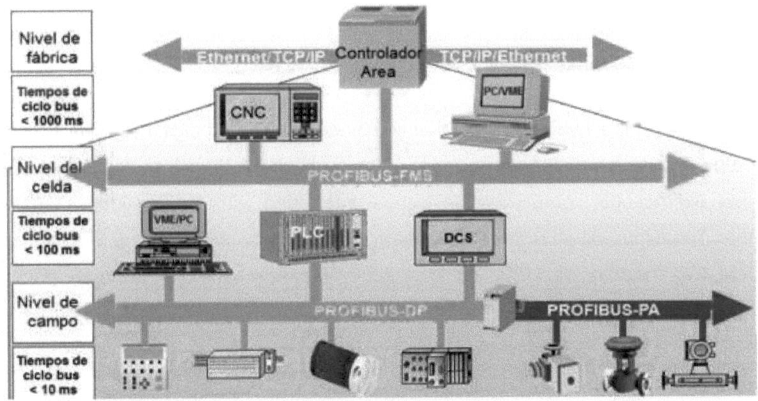

Figure N° 9Profibus network

1.4.4.2.2 *ModBus*

It is a protocol developed by Modicon in 1979, used to establish Master-Slave and Client-Server communications between intelligent devices and with field devices. [6]

It defines a message structure that controllers can recognize and use regardless of the type of network they use to communicate. During communications carried out on a Modbus network, the protocol determines how each controller will recognize addresses, whether a message is addressed to it, determine the action to be taken and extract data from the message. In the same way the protocol and response actions are defined.

There are two main variants:

- ModBus RTU: ideal for remote monitoring via radio of field elements (RTU, Remote Terminal Unit), such as those used in water treatment stations, gas or oil installations. It is currently being implemented in sectors outside its original idea, such as home automation or process control (air conditioning, process control, pumping, etc.). It uses serial communication, such as RS-232 or RS-485.

- ModBus TCP/IP: Provides Client/Server communication between devices connected in an Ethernet TCP/IP network.

1.4.4.2.3 Hart

The HART (Highway Addressable Remote Transducer) communication protocol is widely used in the process industry for digital communication with intelligent field instruments.

HART is a hybrid protocol that combines analog and digital signals. It works by superimposing a digital signal on top of the traditional 4-20 mA analog signal. This allows bidirectional communication without interfering with the continuous analog signal. The digital signal is modulated using the frequency shift keying (FSK) method, where two frequencies (1200 Hz and 2200 Hz) represent bits 1 and 0 (Figure 10).Figure N° 10 10).

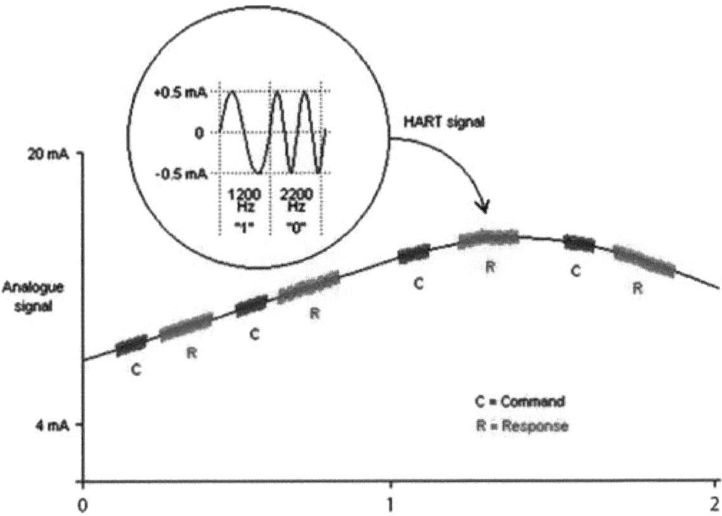

Figure N° 10 10FSK bit coding in the Hart protocol

Each device can transfer up to 256 pieces of data such as: measurement, parameters, status, settings, etc. Power is supplied by the same cable. Up to 15 devices can be connected on the same cable or bus.

Devices based on the Hart protocol are the only ones capable of supporting both analog 4-20mA and digital communication on the same cable, which allows both channels to be used simultaneously to verify the integrity of control loops and enable preventive maintenance in sensitive processes. [6]

1.4.4.2.4 Fieldbus Foundation

The Fieldbus Foundation is the organization dedicated to achieving specifications aimed at creating a single, open fieldbus, as well as hardware and software elements for companies that want to integrate it into their products.

Thanks to digital interoperability between field instruments and multi-vendor systems, it offers the possibility of adding new elements to the control system with the assurance that the bus control functions will not be affected. [6]

It can communicate large volumes of information and provides dedicated control and communication tools for periodic and accurate execution of control functions, eliminating downtime and other communication problems Industrial Ethernet protocols.

1.4.4.3 Industrial Ethernet

Industrial Ethernet is an adapted version of the standard Ethernet protocol used in local area networks (LAN) to meet the needs of industrial automation and control applications. This adaptation includes improvements in robustness, reliability, determinism and security, making it suitable for harsh industrial environments and critical applications.

The use of PCs, LANs and the Internet provides a wealth of services and products from which industry can benefit. Basically, some of these benefits are: the use of a company's existing Ethernet cabling infrastructure, which significantly reduces installation costs, easy access to the network using a PC connected directly to the Internet, the reasonable cost of Ethernet interconnection technology, etc. [7]

1.4.4.3.1 Ethernet/IP

Ethernet/IP (EIP) is a high-level application layer protocol developed for the industrial automation environment. It works with the TCP/IP protocol stack, using all traditional Ethernet hardware and software for the configuration, access and control of industrial automation devices. Ethernet/IP classifies Ethernet nodes as predefined device types with specific characteristics.

It uses all the traditional transport and control protocols in conventional Ethernet, including TCP, IP and media access found in existing network interfaces and devices on the market. [7]

1.4.4.3.2 Profinet

Profinet (Process Field Net) is an industrial communication standard developed by Siemens and managed by the PROFIBUS & PROFINET International (PI) organization. It is based on Ethernet and designed for industrial automation.

It allows the integration of fieldbuses (in particular PROFIBUS) in a simple way and without any modification. In this way, standardized and established IT (information technology) techniques in the area of office automation can also be used in the world of automation, making it possible to link the enterprise resource planning level with the production level and the field level. [8]

With the increasing adoption of the Internet of Things (IoT) and Industry 4.0, Profinet is incorporating advanced cybersecurity features, greater integration with IT systems, and improvements in speed and performance to support more demanding applications.

1.4.4.4 Protocols in wireless systems

Wireless networks are networks that use radio waves to connect devices without the need for cables of any kind, providing flexibility and reducing installation and maintenance costs.

There are several systems, which are distinguished from each other according to their scope or capacity, each being suitable for different tasks.

According to the area of application and signal range, they can be classified as shown in the following figure Figure N° 11.

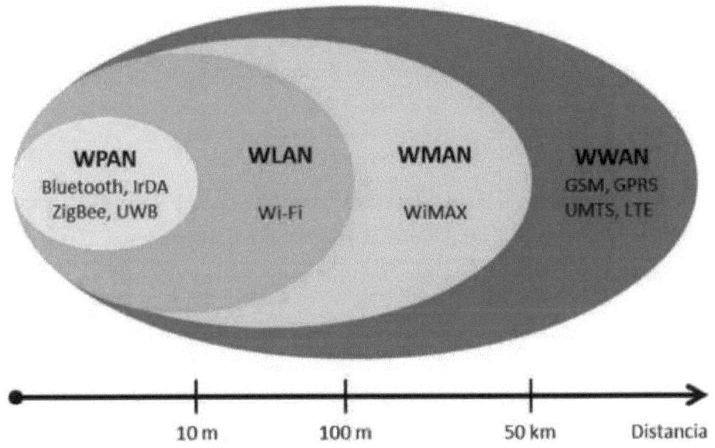

Figure N° 11Classification of wireless networks [9]

The most widespread wireless protocols in the field of industrial automation and data acquisition will be analyzed.

1.4.4.4.1 Wi-Fi (IEEE 802.11)

Wi-Fi is a technology based on IEEE 802.11 standards that uses radio frequencies (RF) to extend Ethernet-based wired local area networks (LANs) to Wi-Fi enabled devices, allowing devices to receive and send information from the Internet.

It uses the Internet Protocol (IP) to communicate between endpoint devices and the LAN. A Wi-Fi connection is established using a wireless access point that is connected to the network and allows devices to access the Internet.

1.4.4.4.2 Bluetooth

Bluetooth is a standard for short-range wireless interconnection of cell phones, computers and other electronic devices.

It sends and receives radio waves in a band of 79 different frequencies (channels) centered at 2.45 GHz, isolated from radio, television and cellular phones, and is reserved for use by industrial, scientific and medical devices. Bluetooth short-range transmitters have very low power consumption and are more secure than wireless networks that operate over longer ranges, such as Wi-Fi.

1.4.4.4.3 ZigBee

ZigBee is a 2.4 GHz mesh local area network (LAN) protocol. It was developed as a specification based on IEEE 802.15.4 for a set of high-level communication protocols used to create personal area networks with small, low-power digital radios.

ZigBee devices transmit data over long distances by passing it through a mesh network of intermediate devices to reach the most distant ones, at a defined rate of 250 kbps.

It is typically used in low data rate applications that require high scalability, long battery life and secure networks. It is simpler and less expensive than Bluetooth or Wi-Fi and is commonly used for home, building and industrial automation applications, such as controlled lighting and thermostats, home energy monitors, smart metering, medical device data collection, traffic management systems and for low bandwidth applications. [9]

1.4.4.4.4 MQTT

MQTT (Message Queuing Telemetry Transport) is a lightweight messaging protocol designed for machine-to-machine (M2M) communication and the Internet of Things (IoT). It is especially suitable for applications where bandwidth is limited, the network is unreliable or devices have limited resources.

The MQTT protocol has become a standard for IoT data transmission as it offers the following benefits:

- **Lightweight and efficient**: The implementation of MQTT in the IoT device requires minimal resources, so it can be used even in small microcontrollers. For example, a minimal MQTT control message can have as little as two bytes of data. MQTT message headers are also small in order to optimize network bandwidth.

- **Scalable**: The implementation of MQTT requires a minimal amount of code that consumes very little power in operations. The protocol also has built-in features to support communication with a large number of IoT devices. Therefore, you can implement the MQTT protocol to connect to millions of these devices.

- **Reliable:** Many IoT devices connect over unreliable cellular networks with low bandwidth and high latency. MQTT has built-in features that reduce the time it takes for the IoT device to reconnect to the cloud. It also defines three different QoS levels to ensure reliability for IoT use cases: at most once (0), at least once (1) and exactly once (2).

- **Secure**: MQTT makes it easy for developers to encrypt messages and authenticate devices and users using modern authentication protocols such as OAuth, TLS1.3, client-managed certificates, etc.

- **Supported**: Several languages, such as Python and C++, have extensive support for the implementation of the MQTT protocol. Therefore, developers can quickly implement it with minimal coding in any type of application.

Principle of operation of MQTT:

The MQTT protocol operates on the principles of the publish or subscribe model. In traditional network communication, clients and servers communicate directly with each other. Clients request resources or data from the server, and the server processes and sends a response. However, MQTT uses a publish or subscribe pattern to decouple the message sender (publisher) from the message receiver (subscriber). A third component, called a message broker, controls the communication between publishers and subscribers. The broker's job is to filter all incoming messages from publishers and distribute them correctly to subscribers. The broker decouples publishers and subscribers as follows:

Spatial decoupling: The publisher and subscriber do not know each other's network location and do not exchange information such as IP addresses or port numbers.

Time decoupling: The publisher and the subscriber do not run or have network connectivity at the same time.

Synchronization decoupling: Both publishers and subscribers can send or receive messages without interrupting each other. For example, the subscriber does not have to wait for the publisher to send a message. [10]

1.4.4.5 Communication protocols for Arduino

In the Arduino domain, several communication protocols are used that allow interaction between the microcontroller and other devices, sensors, actuators and systems. Some of the most common protocols used with Arduino are described below:

1.4.4.5.1 Serial (UART)

The UART (Universal Asynchronous Receiver-Transmitter) communication protocol is a synchronous, serial communication protocol that uses two ports to send and receive information, Tx and Rx, respectively. [12]

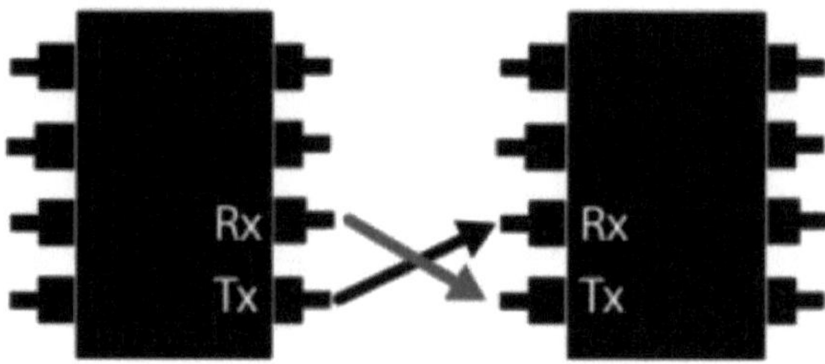

Figure N° 12 12UART ports connection [12]

Asynchronous means that there is no clock signal that coordinates both the sender and the receiver. These are coordinated using start and stop bits present in the message to be sent. For two devices communicating via UART to understand each other, it is necessary that the communication protocol they are using has the same configuration in both devices, so that they can encode and decode the transmitted data. The configuration parameters are as follows:

- Data bits: the number of data bits in a data packet. They represent the information to be transmitted (numbers, characters, instructions, etc.).
- Start and stop bits: bits used to specify the start and end of a data packet.

- Parity bit: The parity bit is used to detect errors in the data packet. In case the receiver detects (thanks to the parity bit) that a data packet contains errors, it asks the transmitter to send that data packet again.

- Data transmission rate: this value is specified in baud per second, one baud being one transmitted symbol. Depending on the modulation scheme, a symbol can represent one or more bits of information. In the case of Arduino, it is true that 1 baud = 1 bit.

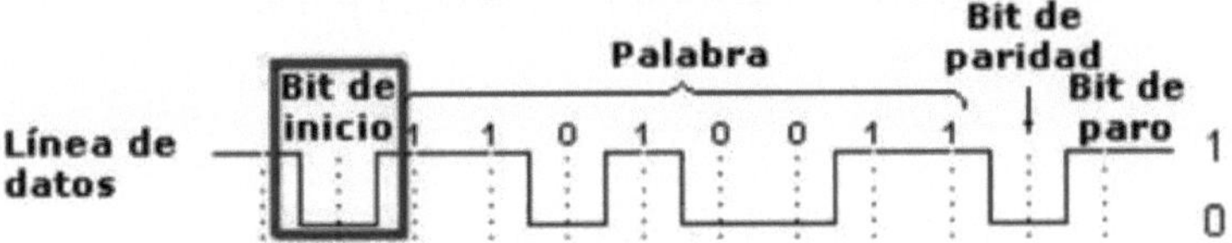

Figure N° 13Data packet of a serial transmission

The most common configuration is 8N1 (8 data bits, no parity and 1 stop bit). This is the standard configuration used by the Arduino.

1.4.4.5.2 I2C protocol

The I2C (Inter-Integrated Circuit) protocol is a serial and synchronous communication protocol. It is also known as TWI (Two Wired Interface). This protocol works with a master-slave architecture, i.e. in this type of architecture there are two types of devices:

- Master: It is the one that initiates and coordinates the communication. The master is responsible for establishing communication with each of the slaves either to send or receive data. When working with I2C with Arduino, it is usually the Arduino board that operates as the master.

- Slave: Slaves are waiting for a master to communicate with them, either to receive data or to send data. Generally, it is the peripherals, sensors and actuators that work as slaves in an I2C communication, although it is also usual that a microcontroller has to operate as a slave.

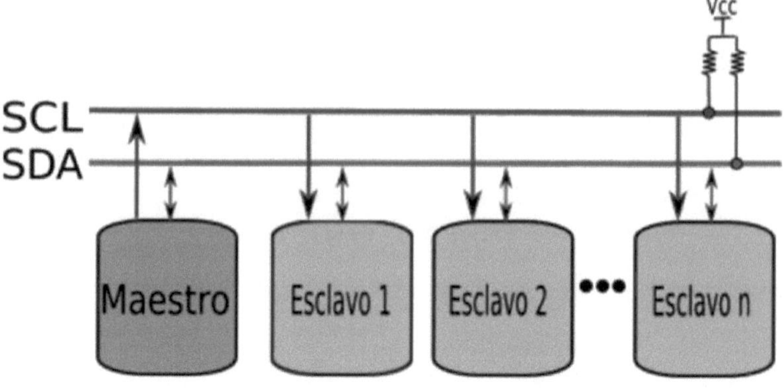

Figure N° 14 14I2C communication

As shown in the figure, only 2 wires are needed in I2C:

- SDA (Serial Data): It is where the serial information travels.
- SCL (Serial Clock): This is the clock signal that synchronizes the communication.

In I2C, each slave is assigned a unique address. The master uses that address to send or receive information to or from that particular slave. In Figure Figure N° 15 shows how an I2C communication message is composed:

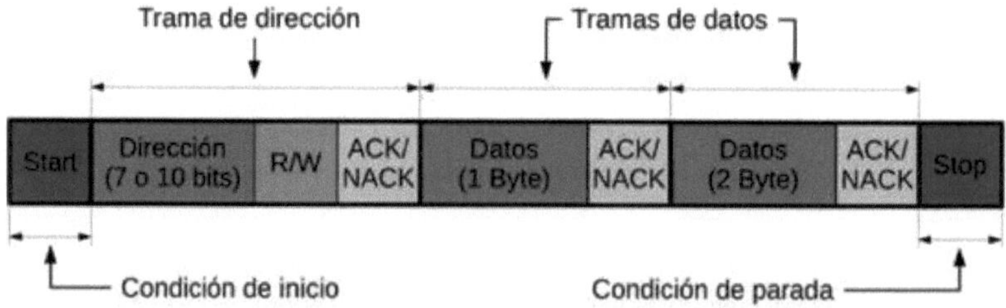

Figure N° 15Composition of an I2C message

The message is composed as follows:

- The start condition (Start) makes the slave devices aware that a communication is about to be initiated.

- The address frame indicates the address of the slave with which communication is to be established.
- The R/W bit indicates whether you want to send information from the master to the slave or vice versa.
- Data frames contain the data (in binary language) to be transmitted.
- The ACK/NACK bits (acknowledgement bits) are sent by the device receiving the data to indicate to the sender that it has received them. In the case of the address frame, the slave uses the ACK bit to indicate to the master that it is the slave with which it wants to communicate.
- The Stop bit is used to indicate that the communication is terminated and you want to release the SDA bus.

The I2C communication protocol is a synchronous communication protocol, which implies that there is a line common to all devices through which the clock signal circulates. In the case of I2C this line is SCL.

The clock signal is a pulse train, when this signal reaches a "high" value both communicating devices read the status of the data signal on the SDA line. [11]

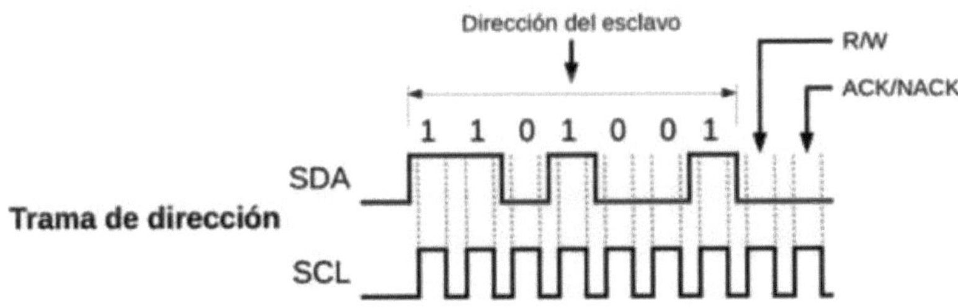

Figure N° 16 16I2C Signal synchronization

1.4.4.5.3 SPI Protocol

The SPI (Serial Peripheral Interface) communication protocol is a four-wire synchronous serial communication protocol that employs a master-slave architecture as shown in Figure 17. Figure N° 17.

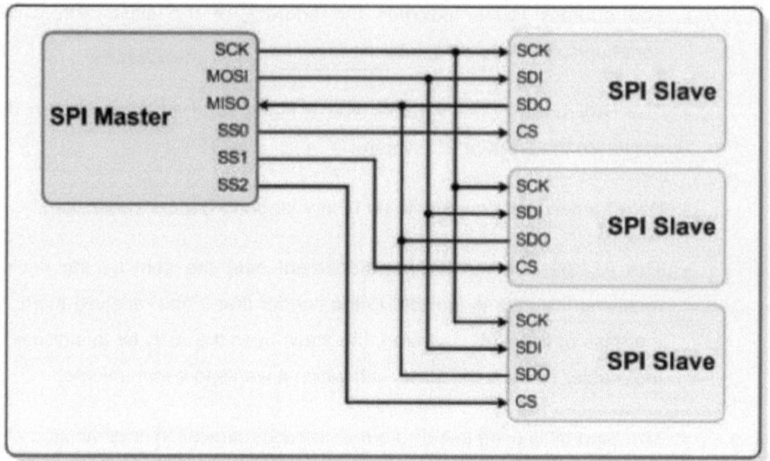

Figure N° 17SPI Communication

Unlike I2C, in SPI each slave has its own slave selection line. The meaning of each SPI line is presented below:

- Master Out, Slave In (MOSI): in this line the master sends data to the slave.
- Master In, Slave Out (MISO): in this line the slave sends data to the master.
- Serial Clock (SCK): is the clock signal that synchronizes the communication.
- Slave Select (SS): this line tells a particular slave to become active, either to send data or to receive data.

One difference with respect to I2C is that SPI has two data wires (MOSI and MISO), so both master and slaves can send and receive data at the same time: it is a full-duplex communication. [11]

1.4.5 Open Source and Open Hardware

The term "open source" refers to an approach to software development that promotes accessibility, transparency and collaboration. In an open source project, the source code of the software is publicly available for viewing, modification and redistribution by anyone.

Open source software has the advantage of being able to be developed in a decentralized and collaborative way, being nourished by peer review and community production. In addition, it is usually more economical, flexible and durable than its proprietary alternatives, since the communities are in charge of its development and not a single author or a single company. [12].

Figure N° 18 18Open source initiative logo

Open source initiative is a non-profit organization dedicated to disseminating and defending the principles and practices of open source software, which defines the terms that must be met for the distribution of open source software. The following are the topics covered in it. [13]:

1. Free Redistribution
2. Source Code
3. Derivative Works
4. Author Source Code Integrity
5. Non-Discrimination Against Persons or Groups
6. Non-Discrimination Against Fields of Action
7. License Distribution
8. License Must Not Be Product Specific
9. License Must Not Restrict Other Software
10. License Must Be Technology Neutral

The term "Open hardware" refers to a similar approach to "Open Source", but applied to the design and manufacture of physical hardware, such as printed circuit boards (PCBs), electronic devices and machines.

Figure N° 19Open source hardware logo

The OSHWA, Open Source Hardware Association, establishes the formal definition of Open Source Hardware, called Statement of Principles 1.0:

Open Source Hardware (OSHW) is hardware whose design is made publicly available for anyone to study, modify, distribute, materialize and sell, both the original and other objects based on that design. The hardware sources (understood as the source files) must be available in an appropriate format so that modifications can be made to them. Ideally, open source hardware uses highly available components and materials, standardized processes, open infrastructure, unrestricted content, and open source tools in order to maximize the ability of individuals to realize and use the hardware. Open source hardware gives freedom to control the technology while sharing knowledge and stimulating commercialization through the open exchange of designs. [14].

The terms that must be met for the distribution of Open Source Hardware are:

1. Documentation
2. Scope
3. Required software
4. Derivative works
5. Free redistribution
6. Attribution

7. Non-discrimination of individuals or groups

8. Non-discrimination to fields of application

9. License distribution

10. The license shall not be specific to a product

11. The license shall not restrict other Hardware or software.

12. The license shall be technology neutral.

1.4.5.1 Advantages of Open Source

- **Transparency and Freedom:** Access to source code allows users to understand how software works, which promotes transparency and the freedom to modify it according to their needs.

- **Collaborative Innovation:** The open development model fosters collaboration between developers around the world, which can lead to greater innovation and the creation of better products.

- **Reduced Cost:** Open source software is generally free or has a significantly lower cost than proprietary software.

- **Flexibility and Customization:** Users have the freedom to modify the software to suit their specific needs.

- **Active Development Community:** Open source projects usually have an active community of developers and users who provide technical support, solve problems and contribute to the ongoing development of the software.

1.4.5.2 Advantages of Open Hardware

- **Access to Designs and Schematics:** Open access to hardware designs and schematics provides insight into how a device works and facilitates design modification and improvement.

- **Manufacturing Freedom:** Users have the freedom to manufacture and assemble their own devices using open hardware designs.

- **Transparency and Trust:** Transparency and trust are promoted by allowing users to inspect and verify designs and components.

- **Collaboration and Community**: Open hardware projects usually have an active community of designers, manufacturers and users who collaborate in the development and continuous improvement of the devices.
- **Customization and Adaptability:** Allows customization and adaptation of devices according to needs.

1.4.6 Arduino

Arduino is an electronics design and prototyping tool that unites "Open Source Hardware" and "Open Source" in a single product. It consists of a printed circuit board with a microcontroller and an integrated development environment (IDE) that allows you to write, load and run programs on the microcontroller [15].

Figure N° 20 20Arduino UNO Development Board

The Arduino microcontroller is usually an AVR family chip from Atmel (now part of Microchip Technology) or the ARM family. The Arduino board includes several digital and analog input/output ports that allow sensors, actuators and other electronic components to be connected.

The Arduino development environment (Figure N° 21 21) provides a library of functions and examples that simplify programming. Arduino programs, called "sketches", are written in a C/C++ based programming language.

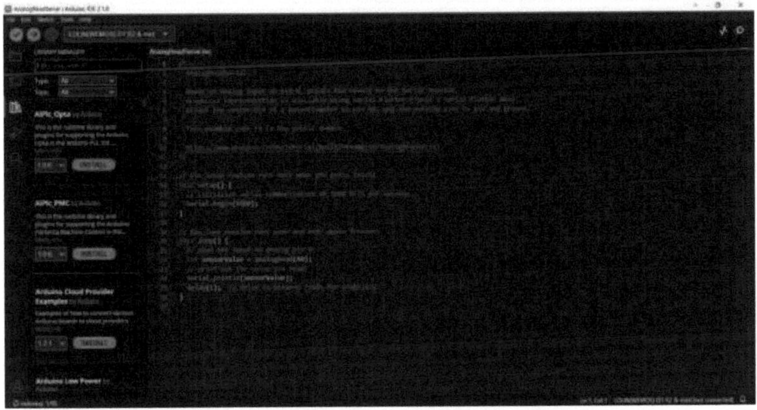

Figure N° 21 21Arduino Integrated Development Environment (IDE)

Its active and collaborative community, along with its extensive documentation and tutorials, make Arduino a powerful tool for students, hobbyists and professionals.

1.4.6.1 Modules for Arduino

Thanks to the great diffusion and popularity of the Arduino development environment, a wide variety of modules have emerged that adapt to it to expand its capabilities. These comprise not only the hardware, but also include the code necessary to use its functionalities. Next, we will see in detail the modules that have been implemented in this project.

1.4.6.1.1 Relay output module

This module provides the ability to command a relay output through an Arduino digital pin, which is very useful for controlling loads higher than our board can handle. It has an optocoupled circuit, which activates the relay coil through a digital signal (depending on the model, it can be activated with a zero or a one). They are commonly available in the market with 1, 2, 4, 8 and up to 16 channels.

To connect, the module is provided with a VCC pin, corresponding to the power supply, a GND corresponding to ground or negative, and the signal pins that activate each of the relays.

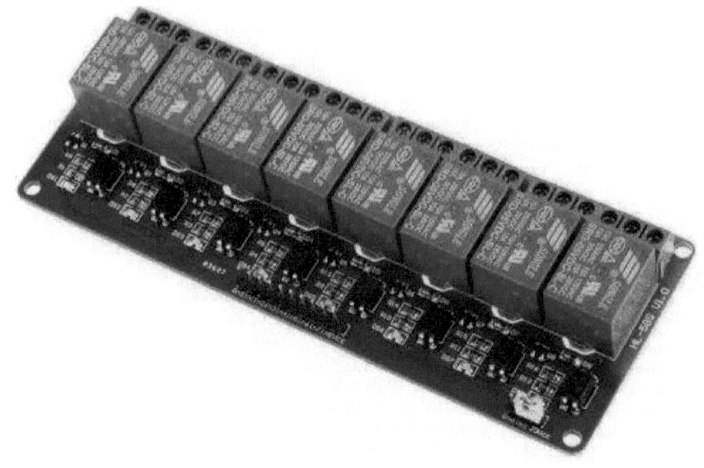

Figure N° 22 228-channel relay module

1.4.6.1.2 LCD display

LCD (Liquid Crystal Display) screens are display devices that allow text information to be displayed in alphanumeric or numeric form.

It is composed of a matrix of pixels that can display text, numbers and graphics. The resolution and size of the display may vary according to the model, the most common being the 16×2 LCD (2 rows and 16 characters) (Figure N° 23), 20×4, 20×2 and 40×2. They come with an LED backlight that allows viewing the display in low light environments, which can be blue, yellow or green.

Figure N° 23LCD display 1602

LCD displays can be connected to the Arduino through different communication interfaces, such as parallel, serial (SPI) or I2C. The I2C interface is commonly used as it requires fewer pins and makes connection easier, but it requires the addition of an I2C interface module such as the one shown in the Figure N° 24 24. This module allows communicating with the display using only 4 pins (GND, VCC, SDA and SCL). It incorporates a potentiometer to regulate the contrast and sharpness of the characters, and pins A0, A1 and A2 which are used to set the device address on the I2C bus.

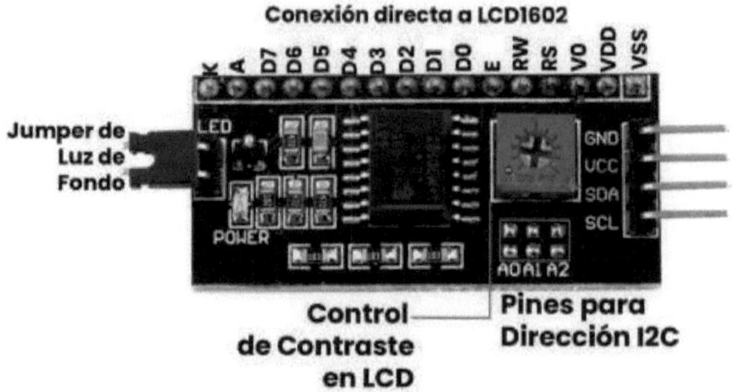

Figure N° 24 24I2C interface module

1.4.6.1.3 Voltage regulator LM2596

The LM2596S circuit (Figure N° 25) is an integrated circuit [16] that provides the capability to regulate or step down the input voltage of the circuit. The integrated circuit handles an operating range of 1.23V to 40V and the output voltage is adjustable through a precision potentiometer.

Figure N° 25Voltage regulator LM2596

Table No. 1 1. LM2596 Voltage Regulator Specifications

Category	Specification
Operating voltage	4.0V ~ 40V DC
Output voltage	1.23V ~ 37V DC adjustable (input voltage must be at least 1.5V more than output)
Output current	max. 3A, 2.5A recommended (use heatsink for currents higher than 2A)
Power output	50-70W
Conversion efficiency	0,92
Load regulation	S (I) ≤ 0.8%.
Voltage regulation	S (u) ≤ 0.8%.
Frequency of work	150KHz
Ripple at the exit	30mV (max.), 20M bandwidth
Working temperature	-40°C ~ +85°C
Protection	Short circuit and overtemperature
Dimensions	4.2cm x 2.3cm x 1.2cm
Weight	11 grams

1.4.6.1.4 Optocoupler PC817

An optocoupler, also called optoisolator or optically coupled isolator, is a transmitter and receiver device that functions as a switch activated by the light emitted by an LED diode saturating an optoelectronic component, usually in the form of a phototransistor or phototriac. In this way, a photoemitter and a photoreceptor are combined in a single semiconductor device, the connection between them being optical. These elements are housed in an encapsulation, usually of the DIP type, and are often used to electrically isolate very sensitive devices. [17].

In the case of the PC817, it is a general-purpose optocoupler in a single-channel DIP-4 package (Figure N° 26 26).

PINOUT
Optoacoplador PC817

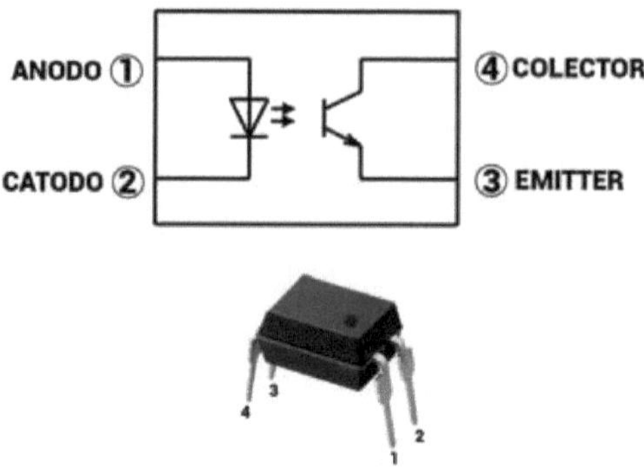

Figure N° 26 26Optocoupler PC817

The most common applications for the PC817 Optocoupler are:

- I/O isolation for MCUs (microcontroller units)
- Noise suppression in switching circuits
- Signal transmission between circuits of different potentials and impedances.

1.4.6.1.5 Darlington transistor array ULN2803

The ULN2803 integrated circuit (Figure N° 27 27) is a high-voltage, high-current Darlington transistor array. The device consists of eight NPN Darlington pairs that provide high voltage outputs with common cathode protection diodes for switching inductive loads. The rated collector current of each Darlington pair is 500 mA. Darlington pairs can be connected in parallel for higher current capability.

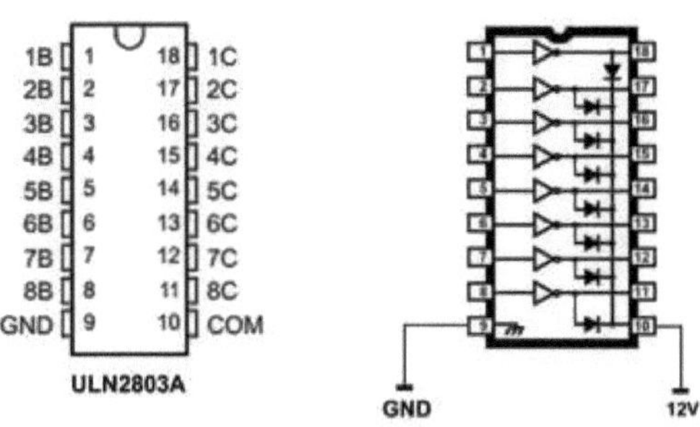

Figure N° 27 27Darlington Array ULN2803

A Darlington pair (Figure N° 28 28) is an arrangement of two bipolar transistors that are used together to provide high current gain. The Darlington configuration allows one transistor to control the current of another transistor, resulting in a combined current gain much higher than that of a single transistor.

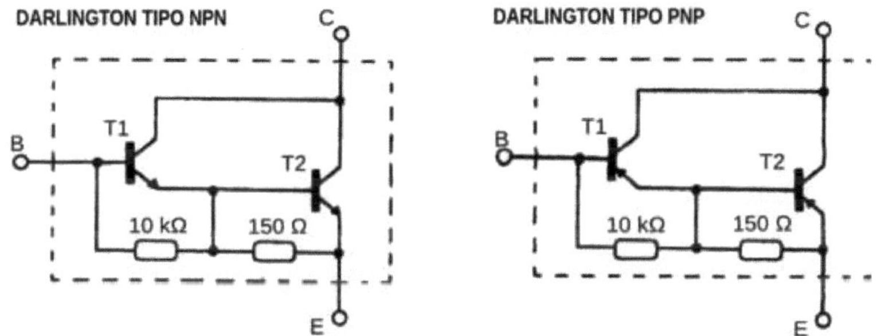

Figure N° 28 28Darlington Pair

Incorporating this integrated circuit into the PLC design significantly increases the microcontroller's ability to handle loads. While an Arduino board can provide up to 20 mA of current on each pin, the inclusion of this circuit allows a capacity of up to 500 mA per pin.

1.4.6.2 Sensors

Sensors are devices that transform physical variables from the environment into electrical variables that can be read and interpreted by a microcontroller. They provide electronic systems with information from the world around them, allowing them to make decisions, control processes or take actions in response to detected changes.

Some of the most commonly used sensors within the Arduino development environment are:

- Temperature sensor
- Humidity sensor
- Proximity sensor
- Hall effect sensor (flow meter)
- Pressure sensor
- Total dissolved solids sensor
- Color sensor
- Sound sensor
- Linear and angular motion sensor
- Level sensor

The following is a detail of the sensors that are involved in the control of the water treatment equipment.

1.4.6.2.1 Hall effect sensor - Flow meter

A flow meter is a sensor used to measure the flow rate of a fluid through a pipe. There are different methods to perform this measurement, but the most common and cheapest available for the Arduino environment is the Hall effect sensor. In this project, the YF-B5 flow sensor (Figure N° 29), which consists of a brass tube with a propeller inside and a Hall effect sensor on the outside. A small magnet is attached to one of the propeller blades. When fluid flows through the tube and rotates the propeller, the magnet also rotates, generating a variable magnetic field. As the propeller rotates and the magnet passes close to the sensor, the varying magnetic field induces a voltage in the Hall effect sensor. This voltage is proportional to the rotational speed of the propeller, which is used to calculate the fluid flow rate.

The sensor output is a square wave whose frequency is proportional to the flow through it:

$$f(Hz) = K.Q\left(\frac{l}{min}\right) \Rightarrow Q\left(\frac{l}{\min}\right) = \frac{f(Hz)}{K}$$

The conversion K-factor between frequency (Hz) and flow rate (L/min) depends on the construction parameters of the sensor. The manufacturer provides a reference value in its Datasheet. However, the K constant depends on each flowmeter. With the reference value we can have an accuracy of +-10%. If a higher accuracy is required, a test must be performed to calibrate the flowmeter.

Figure N° 29Flow sensor YF-B5

1.4.6.2.2 Total Dissolved Solids Sensor - Conductivity Meter

A TDS (total dissolved solids) meter measures the amount of total dissolved solids such as salts, minerals and metals in water. As the amount of dissolved solids in the water increases, the conductivity of the water increases, allowing the total dissolved solids to be calculated in ppm (mg/L).

The principle of operation of a TDS meter is based on the electrical conductivity of water. When dissolved solids are present in water, these ions or charged particles can conduct electric current. Therefore, the higher the concentration of dissolved solids in water, the higher its electrical conductivity.

The TDS meter (Figure N° 30 30) uses electrodes immersed in water to measure its electrical conductivity. When the electrodes are in contact with the water, a small electric current is generated between them. The sensor measures this current and converts it into a TDS reading in ppm (parts per million).

Figure N° 30 30TDS sensor KS0429 keyestudio V1.0

1.4.6.2.3 *Float level switch*

These switches consist of a casing made of a material with a density lower than that of the fluid, which allows for flotation, and a ball (Figure N° 32) that moves freely inside it. When the liquid level drops, the switch begins to hang from its signal cable and its angle of inclination changes. This change of inclination causes the ball to roll in one direction or the other, actuating or interrupting an internal contact.

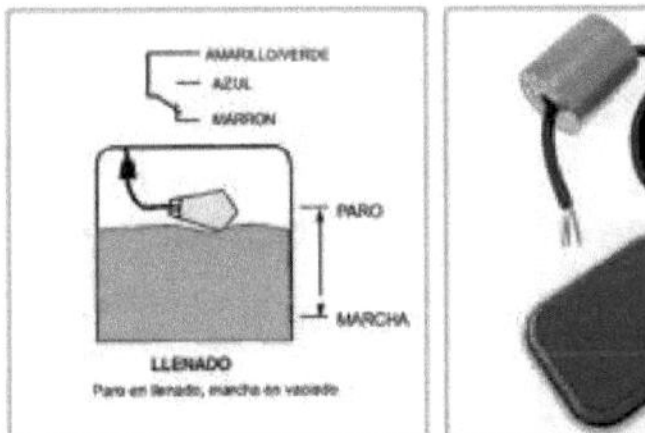

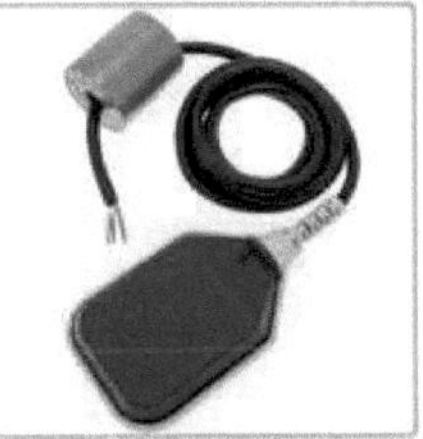

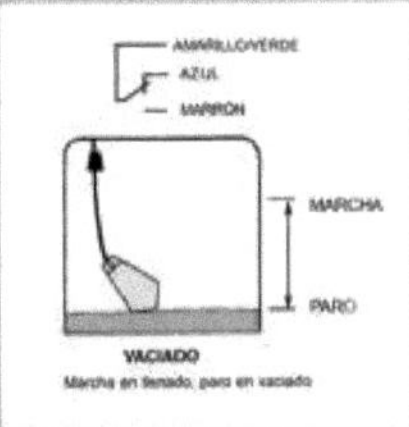

Figure N° 31Float level switch

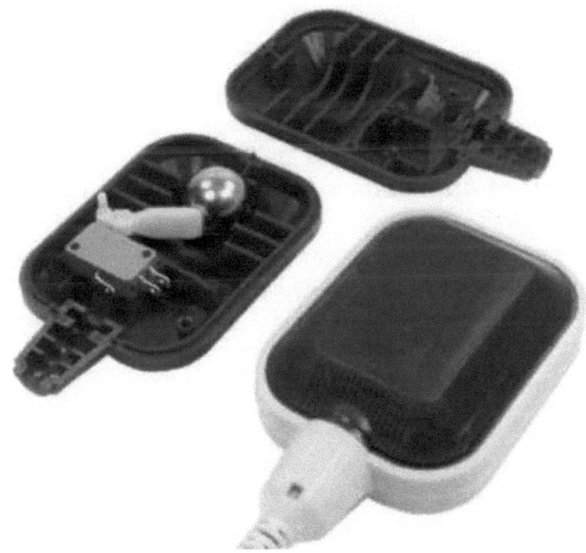

Figure N° 32Interior of the float level switch

1.4.6.2.4 *Pressure switch*

The pressure switch is a device that monitors the pressure of a liquid or gas in a system. Its main function is to detect when the pressure reaches a predetermined value and, in response, activate or deactivate an electrical circuit.

Figure N° 33Pressure switch

It has a connection at the bottom to connect the pressure switch to the pressurized system, and two inputs for the electrical connections. When the cover is removed, there are two bellows pressed with nuts, through which it is possible to adjust the operating pressures of the electrical contacts.

1.4.7 Context of Application: Water Treatment Plant

This project has been conceived with the objective of automating the operation of a water treatment plant, which purifies water to obtain the appropriate parameters for the irrigation of a greenhouse.

1.4.7.1 Description of the Water Treatment Plant

In the block diagram (Figure N° 34 34) shows the different elements that are part of the water treatment system and their relationship with each other.

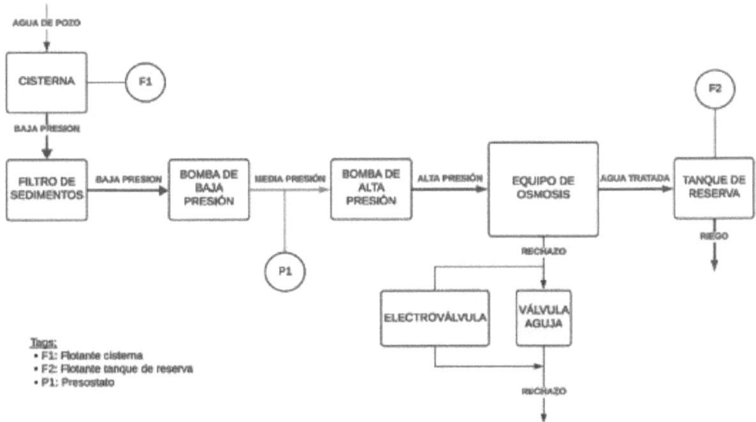

Figure N° 34 34Block diagram: Water treatment plant

Cistern: it is the reservoir of raw water from the groundwater, coming from a borehole.

Sediment filter: It is a basket type filter, with a mesh for the filtering of solids of large granulometry.

Low pressure pump (booster): It is the pump in charge of supplying the adequate pressure for the high pressure pump, due to the fact that the latter has a higher suction pressure than the one available by the potential energy of the tank tank at height.

High pressure pump: It is the pump in charge of generating the necessary pressure for the reverse osmosis effect to occur, and the purified water can pass through the membranes.

Osmosis equipment: Consists of a tube with osmosis membranes inside. These membranes are semi-permeable, with microscopic pores that allow water to pass through them, but block the passage of larger contaminants. High pressure is required for this separation to occur. The purified water passes through one conduit to the holding tank, and the contaminants are separated on the other side, known as reject.

Needle valve: This valve is manually regulated to restrict the rejection flow line in order to increase the pressure inside the membrane tube of the osmosis equipment.

Solenoid valve: This electrically actuated valve remains closed during normal operation of the equipment, and opens during the equipment shutdown process. When fully opened, it allows the unrestricted passage of the rejection, which generates a greater water circulation, thus dragging all the impurities that may be deposited on the osmosis membrane faces.

Reserve tank: It is a reservoir where purified water is stored, available for irrigation.

1.4.7.2 Automation requirements

The following are the elements involved in automation, which have a direct influence on PLC decisions and actions.

1.4.7.2.1 Tickets

- Floating tank: float level switch.
- Floating tank: float level switch.
- Pressure switch: measures the pressure generated by the low pressure pump and closes a contact when the set pressure is reached.

1.4.7.2.2 Exits

- Low pressure pump
- High pressure pump
- Solenoid valve: It is a valve that opens when current is supplied through its solenoid. It is in charge of opening the pipe to backwash the osmosis membranes.
- Alarm light: Red light that is mounted on the front of the electro board, and is activated when a problem occurs in the starting process of the equipment.

1.4.7.2.3 Control philosophy

Equipment start-up: The equipment starts when the start-up permissives are activated. First the low pressure pump is turned on. After 20 seconds, if the pressure switch indicates that the pressure has reached the set value, the high pressure pump is turned on. This time delay in starting the high pressure pump is to ensure that the entire pipeline is filled with water at a lower pressure to avoid water hammer.

Start-up permits:

- High cistern float: indicates that there is water for pump suction.
- Low reservoir tank float: The treated water reservoir tank is low and needs to be filled.

Equipment shutdown: When any of the start-up permissives is deactivated, the equipment shutdown process is initiated. This consists of opening the solenoid valve, which causes the osmosis membranes to be flushed, and 60 seconds later, turning off the high and low pressure pump.

Alarm - failed start: To avoid false readings on the pressure switch caused by starting the low pressure pump, the low pressure pump is turned on and waits for 4 seconds to consider its reading. If after this time the pressure does not reach the set value, the low pressure pump is turned off and a new start is attempted (if the start-up permissives allow it). If after 3 consecutive attempts, the start is not executed, the low pressure pump is turned off and an alarm is activated indicating "Start failed".

1.4.7.3 System implementation

In the Figure N° 35 shows the real osmosis equipment that has been automated using the open source PLC developed. In Figure Figure N° 36 shows each of the components of the process indicated in Figure No. 34. Figure N° 34 34Block diagram: Water treatment plant.

In the Figure N° 37 shows in greater detail the PLC mounted on the electrical panel of the osmosis equipment.

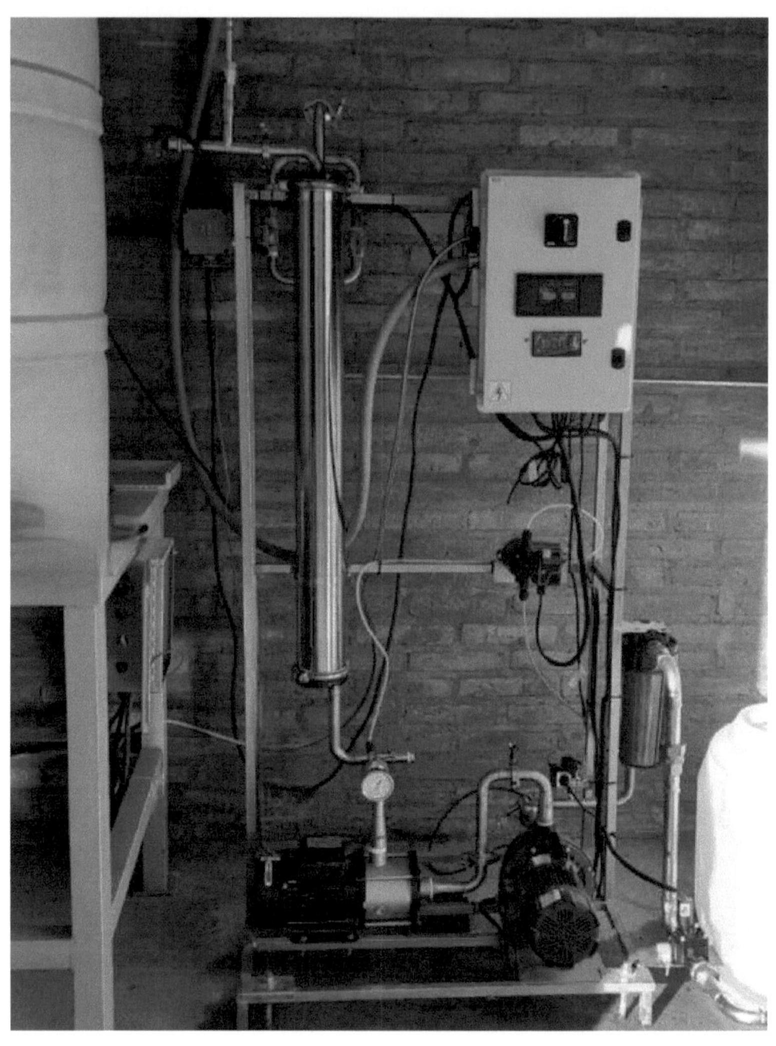

Figure N° 35Osmosis equipment - actual implementation

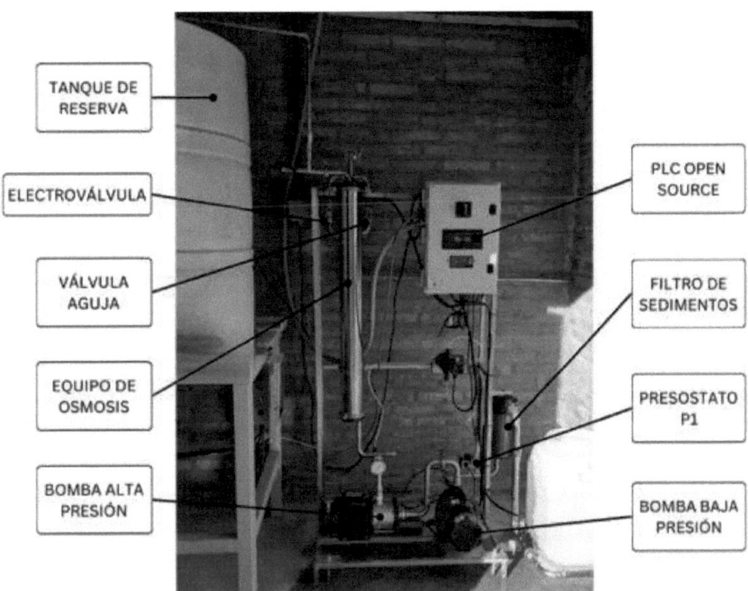

Figure N° 36Components of osmosis equipment

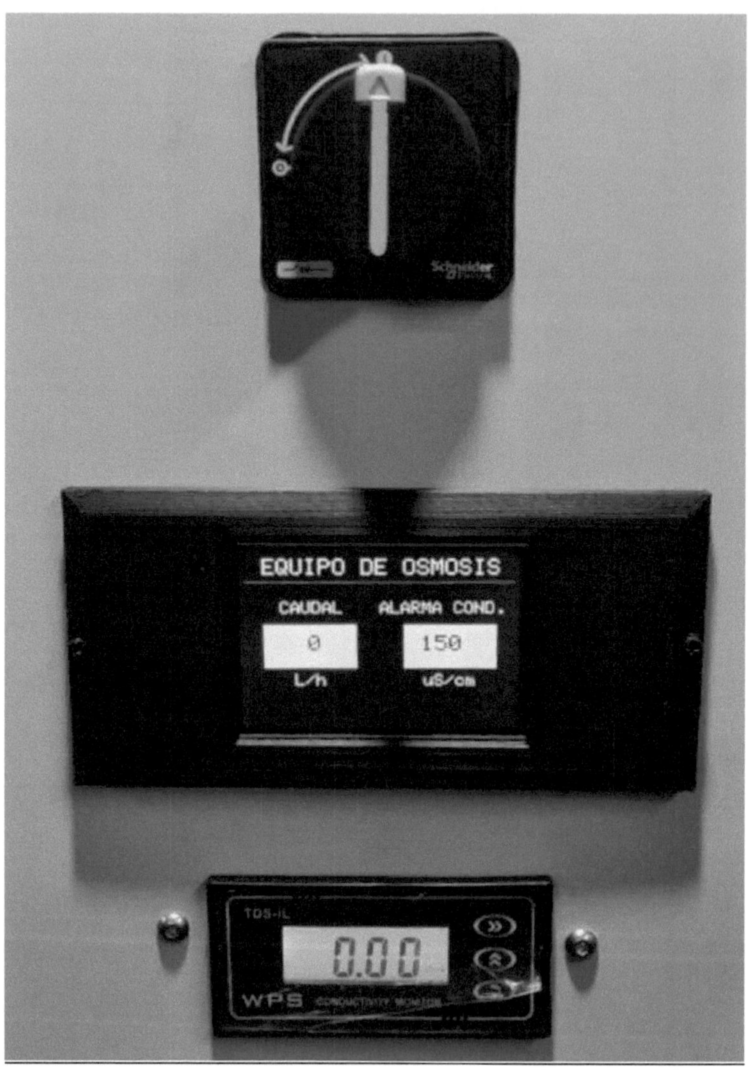

Figure N° 37PLC open source - actual implementation

1.4.8 Justification

When carrying out an automation project, we can find in the local market a wide variety of industrial PLCs, which cover a wide range of functions, but are available at a high cost. With a low budget, the possibility of automation is lost, or we must resort to the development of solutions using open source microcontrollers. This option has the advantage of being cheaper, but it represents a lot of work time and the result is not always the most robust. For this reason, this project seeks to provide end users with an affordable solution, with basic functions, but that meets the performance in a reliable and satisfactory manner, so that they can focus their resources and energy on the automation project itself, improving their processes effectively and without compromising the quality and stability of the implemented system.

1.4.9 State of the art

1.4.9.1 Industrial PLCs

1.4.9.1.1 Siemens S7-1200 PLC

The PLC S7-1200 is a programmable logic controller developed by the German company Siemens, designed for industrial automation, which stands out for its flexibility, ease of use and ability to integrate with other systems.

The main characteristics are as follows:

- Modularity: It is designed as a modular system, so expansion modules can be added to increase its capabilities as required. These modules can be either analog or digital inputs or outputs, communication modules, technology modules (fast counters, positioning, energy measurement, etc.) and function modules.

- Communication capabilities: Supports multiple communication protocols such as PROFINET, PROFIBUS, Modbus TCP/IP, and point-to-point communication. It has an integrated Ethernet port, which facilitates integration into industrial networks.

- Programming: It is programmed using TIA Portal (Totally Integrated Automation) software, which provides an integrated and easy-to-use development environment. It supports various programming languages such as ladder diagram, function block diagram, structured text, etc.

- Advanced functions: Integrates PID control functions, enabling accurate and efficient closed-loop controllers, and supports the creation of web applications for monitoring and control from web browsers.

- Security: Provides security features to protect data and system access, such as user authentication and data encryption.

Figure N° 38Siemens S7-1200 PLC

Table No. 2 2Characteristics of S7-1200 models [19]

Model	Memory	Inputs/Outputs	Communication Interfaces	Additional Modules
S7-1211C	25 KB	6 DI / 4 DO	1 x Ethernet	Up to 2
S7-1212C	50 KB	8 DI / 6 DO	1 x Ethernet	Up to 3
S7-1214C	100 KB	14 DI / 10 DO	1 x Ethernet, 1 x RS-485	Up to 8
S7-1215C	125 KB	14 DI / 10 DO	2 x Ethernet, 1 x RS-485	Up to 8
S7-1217C	125 KB	14 DI / 10 DO	2 x Ethernet, 1 x RS-485	Up to 8

The Table No. 2 2 shows different models of the S7-1200 PLC with their main characteristics.

1.4.9.1.2 Delta PLC TP04P-32TP1R

This Delta Electronics PLC model consists of a programmable logic controller with a 4-line text HMI panel and a touch panel, all integrated in one device.

Figure N° 39Delta PLC TP04P-32TP1R [20]

Table No. 3 3PLC Delta TP04P-32TP1R PLC Characteristics

Feature	Description
Model	TP04P-32TP1R
Digital Inputs	16 DI
Digital Outputs	16 DO (Relay)
Expansion Capacity	Allows connection of expansion modules
Communication Interfaces	RS-232/RS-485, Modbus
Power Supply	24 VDC
Display and Keyboard	LCD display and integrated keyboard
Programming	WPLSoft software, supports Ladder Diagram (LD)

1.4.9.2 Industrial PLCs based on Open Source Hardware

Currently, several companies are entering the supply of industrial controllers based on Open Source Hardware, gaining ground in the field of automation. This new generation of controllers offers an innovative and flexible alternative for a wide range of industrial applications and engineering projects, designed to withstand extreme conditions of temperature, humidity and electromagnetic noise, typical in industrial environments.

1.4.9.2.1 Finder Opta - Arduino Pro

The company Finder, an Italian brand present all over the world, which produces electronic and electromechanical components for the civil and industrial sector, has decided to partner with Arduino Pro [18]to launch a new range of Industrial

Smart Relays or Programmable Logic Relays (PLR). This new line is called OPTA (Figure N° 40), and is available in 3 variants. [19]:

- Opta LITE: with Ethernet or ModBus TCP/IP connectivity and USB programming port (type C).
- Opta PLUS: with Ethernet or ModBus TCP/IP connectivity and USB programming port and adds the RS485 connectivity interface.
- Opt for ADVANCED: the most innovative and complete option, equipped with Wi-Fi and Bluetooth.

All three variants feature 12-24 VDC power supply, 8 inputs both digital and analog (0-10 V), as well as 4 relay outputs of 10 A NO contact. Each of them can be obtained from the official Arduino Store website. [20] at a price of U$S146.40, U$D160.80 and U$D193.20 respectively, in the United States or Europe.

It has the great advantage of being programmable both with traditional licensed languages (Ladder, FBD and others) and with free and open source languages (IDE/ARDUINO).

Figure N° 40PLR Finder OPTA

1.4.9.2.2 Industrial Shields

A bit more powerful, we can find the industrial controllers based on Open Source Hardware from Industrial Shields. [21]. This company has developed PLCs based on three different Open Source Hardware options:

PLC Arduino:

In the Figure N° 41 shows one of the models offered by Industrial Shields, which uses original Arduino boards as microcontroller. As general features, they have up to 58 inputs and outputs, which can be analog, digital or relay, handle multiple communication protocols, such as industrial standard, Ethernet, RS232, RS485, LoRa, Narrowband, WiFi, and much more.

They offer the possibility to expand with 127 modules via the I2C system, which means that it can govern up to 7100 I/O in master slave mode, plus additional sensor modules.

Figure N° 41Arduino PLC of Industrial Shields

Raspberry Pi PLC:

Raspberry Pi open source boards are controllers that run an operating system (Linux or Raspberry Pi OS). Implementing this technology in a PLC increases the processing capabilities of the system, allowing to work in real time and with multi-processes, among others.

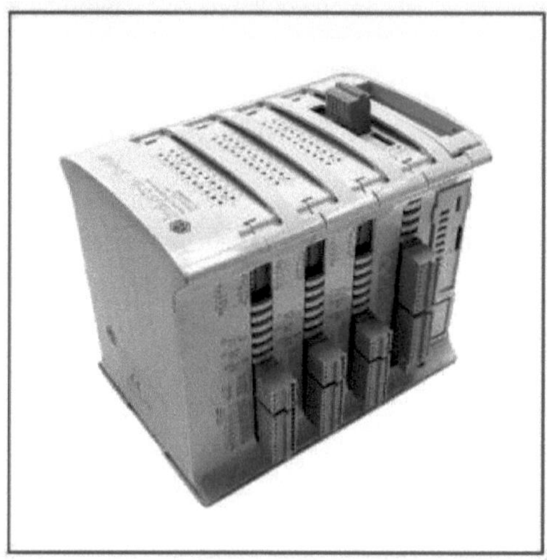

Figure N° 42Industrial Shields' Raspberry Pi PLC

The main advantage of using the Raspberry Pi PLC is that it is compatible with any Integrated Development Environment (IDE) that works with the supported programming languages. This can be accomplished by either accessing an IDE installed on the PLC, or by using an IDE installed on a computer, and then transferring the files. Although a wide variety of programming languages can be used, it is advisable to use Python, C++ and Node-RED, as these are their official languages.

ESP32 PLC:

The ESP32 board-based PLC represents a very similar version of the Arduino-based PLC, but with a higher processing speed. It can be programmed using the Arduino IDE, and is especially suitable for applications with WiFi or Bluetooth connectivity.

Figure N° 43ESP32 based PLC from Industrial Shields

1.4.9.2.3 Controllino

The company Controllino [22] is recognized for manufacturing Arduino-based PLCs, generating a great impact on the automation market. With more than 75 thousand products sold in 150 countries, this company has stood out since its foundation in Austria in 2016.

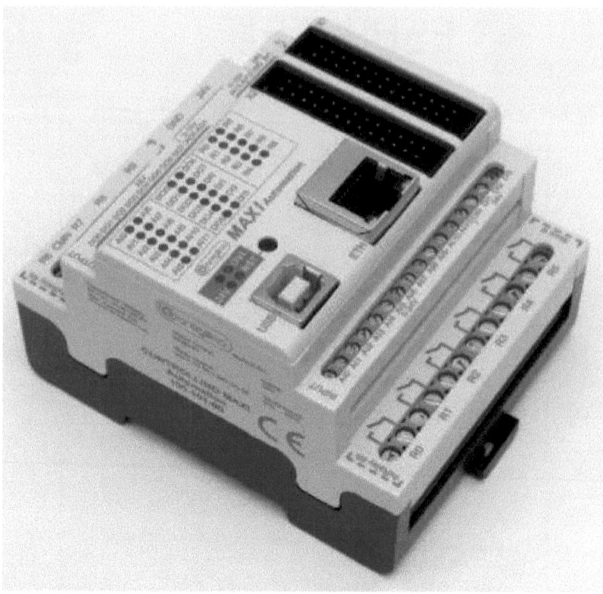

Figure N° 44PLC Controllino

They offer several models, with different numbers of inputs and outputs and communication ports. One of the important features of Controllino PLCs is that their products adhere to the highest industrial and electronic safety standards, such as CE, UL and IEC 61131 certification.

1.4.9.3 Independent open source PLC designs

Thanks to the open source philosophy, many independent PLC design projects have emerged from platforms such as Arduino, in which information and advances are shared among the community, thus generating a momentum of innovation and global growth. Some of the observed designs, which have been solid references for the project in question, are presented below.

1.4.9.3.1 Arduino based PLC from "Electroall".

Electroall [23] is a website that provides information, projects and tutorials about electronics, electricity and industrial automation. It also presents PCB designs for industrial control and energy efficiency.

In the Figure N° 45 you can see the design of its PLC in its latest version V3. It has the following characteristics:

Table No. 4 4. Electroall" PLC Specifications

Specification	Value
Supply voltage	24VDC
Power supply current	65mA
Digital inputs	12-24VDC (8)
Programming environment	Arduino IDE
Minimum environmental conditions	-10°C
Maximum environmental conditions	55°C
RLY Outputs	8
AC output voltage	250V
AC current	5A
DC output voltage	30V
DC current	5A
Dimensions	100x100mm
Recessed	Yes

Figure N° 45PLC with Electroall's Arduino

All the documentation required for free reproduction can be consulted on the website:

- Electronic schematic
- Print Ready Card (JLCPCB)
- Materials and costs
- Complete design in Proteus 8.6

1.4.9.3.2 PLC based on Arduino by "El profe Zurco".

The website of "El profe Zurco". [24] is a blog with a wide variety of electronics and automation projects. Many of them are PLC design projects based on open source technology. In each of them all the necessary documentation for the reproduction of the same is presented, along with informative videos and tutorials that facilitate the understanding, both for its use and for its manufacture.

Figure N° 46PLC of "El profe Zurco".

In the Figure N° 46 shows one of its latest PLC developments based on Arduino (Atmega128A).

1.4.9.4 Open Source Software

Many of the options in the open hardware device market have already been discussed. Now we will show the different open source software options that can be found for PLC programming and for management of devices connected to the Internet.

1.4.9.4.1 Arduino IDE

The Arduino software (IDE), developed by Arduino, is an open source software used to program Arduino boards and many more (NodeMCU, Wemos, Adafruit, Blue Pill, etc). It comprises the environment where the computer code is written and then uploaded to the physical board. It is composed of numerous libraries and a set of examples of simple projects. It is compatible with different operating systems (Windows, Linux, Mac OS X) and supports programming languages (C/C++). It is easy to use for both beginners and advanced users.

The application can be downloaded free of charge [25] or program directly in the online application (Figure N° 47). Recently, the Arduino Cloud environment has been added (Figure N° 48), which provides functionalities to realize IoT projects.

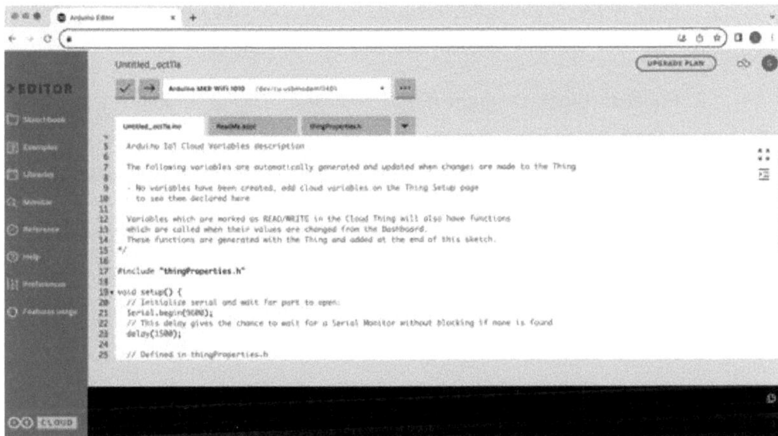

Figure N° 47Arduino IDE Online

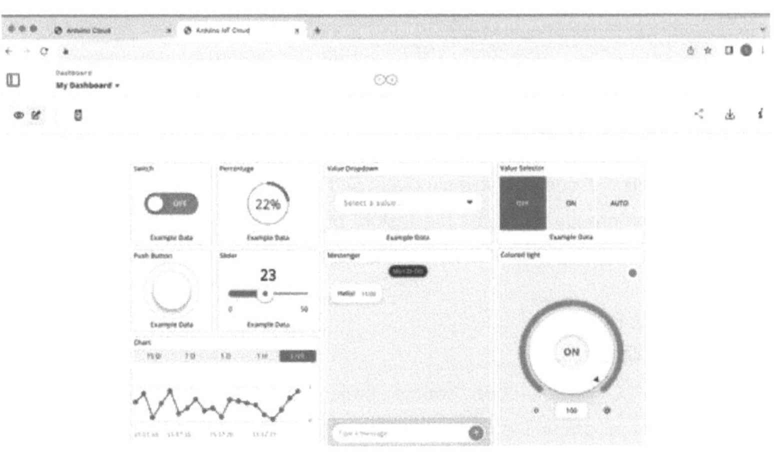

Figure N° 48Arduino Cloud Dashboard

1.4.9.4.2 Codesys

Codesys is a development environment for programming controllers in accordance with the IEC 61131-31 international industrial standard. One of its main peculiarities is that it is not tied to any specific hardware, which makes it possible to program many controllers. It allows programming in 5 different languages:

1. Ladder Diagram Language (LD): This is one of the most common programming languages used in PLC programming. It allows the control logic to be represented using a series of contacts and coils, similar to electrical circuit diagrams.

2. Function Block Diagram (FBD) language: This programming language uses function blocks to represent control logic. It is especially useful for representing complex algorithms and Boolean logic in a visual way.

3. Structured Text (ST): It is a text-based programming language similar to other high-level programming languages such as C or Pascal. It allows writing complex algorithms using control structures such as loops, conditionals and functions.

4. Instruction List Language (IL): This is a low-level programming language that uses a series of instructions to represent the control logic. It is useful for experienced programmers who need precise control over PLC behavior.

5. Sequential Flowchart (SFC): This programming language allows representing the sequence of states and transitions of a control process in the form of a flowchart. It is useful for modeling state-based systems.

In addition, it comes with a simulator and an integrated HMI (Human Machine Interface), which greatly facilitates the task of programming and validating projects. The development environment is free, but the drivers for certain hardware devices are not. [26].

1.4.9.4.3 OpenPLC

OpenPLC is an open source project that provides a programming environment for programming PLCs using open source hardware such as Arduino, Raspberry Pi and other compatible devices. The OpenPLC software can be freely downloaded, used and modified at no charge [27].

It complies with the IEC 61131-3 standard, allowing 5 different programming languages. This allows devices such as Arduino and Raspberry to be programmed in the same way as conventional PLCs.

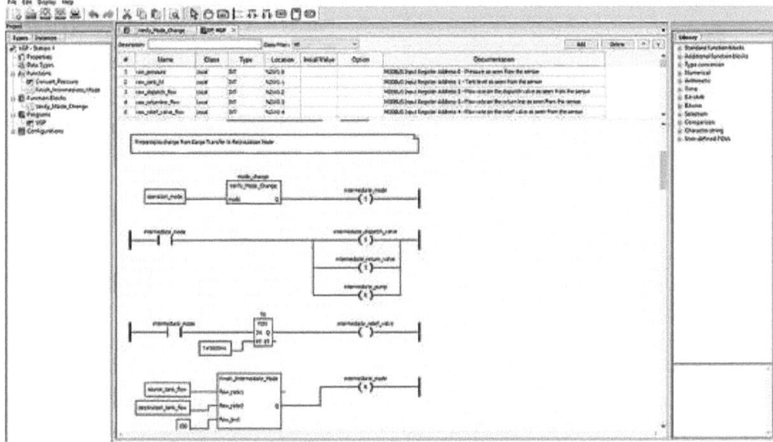

Figure N° 49OpenPLC Editor

1.4.9.4.4 Node-RED

Node-RED is a flow-based development tool designed to facilitate the integration and connection of hardware devices, APIs and online services in a simple way. It works by visually displaying relationships and functions, which makes it possible to program without writing. It is a panel of flows that can incorporate nodes that communicate with each other and can be installed on computers such as Windows, Linux, or cloud servers.

It has become the open-source standard for real-time data processing. It has managed to simplify as much as possible the processes between those who produce information and those who consume it in order to facilitate server-side programming, using visual programming.

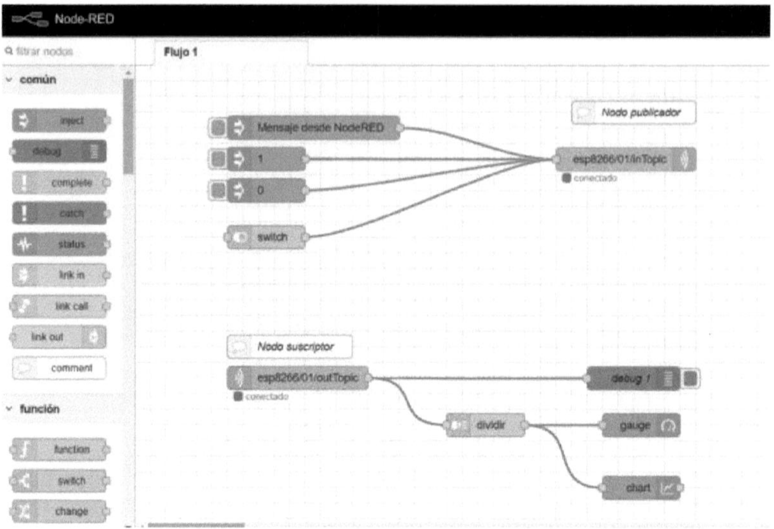

Figure N° 50Node-RED development environment

CAPITULO 2: ANALYSIS AND DEVELOPMENT

The development of this project involved an iterative design process between the different stages that make up the project, before reaching a satisfactory final result.

First, the water treatment system was analyzed (1.4.7 Context of Application: Water Treatment Plant) and the automation needs were defined.

From there, the characteristics and functionalities of the PLC were chosen, taking as a reference the products and systems already existing in the market.

Once the general structure of our PLC was defined, we continued with the design of the electronic circuits and the selection of the required components.

2.1 System architecture

2.1.1 Definition of requirements

The following features were defined for the PLC to meet the control needs of the water treatment system:

Table No. 5 5. PLC Characteristics

Categoría	Especificación
Alimentación	7 V a 40 V (DC)
Entradas	8 entradas digitales 12 V a 24 V (DC)
Salidas	8 salidas a relé 220V – 10 A
Comunicación	Protocolo I2C
Interacción con el usuario	Pantalla LCD para visualización de datos y alarmas
Características adicionales	Sistema de Reloj RTC14 para la gestión y control de la hora y fecha

2.1.2 General structure of the PLC

The different elements that are part of the PLC, and the link between them, are shown in Figure 51. Figure N° 51.

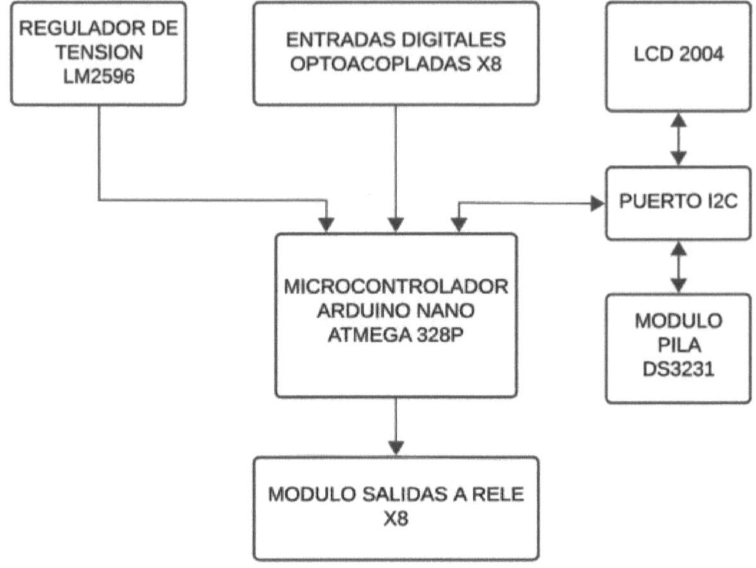

Figure N° 51General structure of the PLC

2.2 Component selection

In order to obtain the characteristics and functionalities required for our system, the following components were selected, corresponding to each of the elements that are part of the system architecture.

2.2.1 Microcontroller

The most important selection is the one that corresponds to the microcontroller, as it is in charge of managing all the other elements. Its capabilities are the ones that determine or limit the functionalities of our PLC.

As responsible for this task, the Arduino Nano development module has been selected (Figure N° 52 52), since it has a sufficient number of inputs and outputs for the requirements of the system, and can be programmed through the free and open source Arduino IDE software (page 66).

Another important point that led to its selection was its low cost and availability in the local market.

Figure N° 52 52Arduino Nano

Table N° 6Technical specifications of Arduino Nano

Category	Specification
Microcontroller	ATmega328P
Operating voltage	5 V
Recommended input voltage	7-12 V
Flash memory	32 KB (2 KB used per bootloader)
SRAM	2 KB
Clock speed	16 MHz
Analog I/O pins	8
EEPROM	1 KB
DC current per I/O pin	40 mA
Digital I/O pins	22
PWM output	6
Power consumption	19 mA
Printed circuit board size	18 x 45 mm
Weight	7 g
Communication port	mini USB type B

2.2.1.1 Arduino Nano Technical Specifications

In the Figure N° 53shows the connections available on the Arduino Nano.

ARDUINO NANO

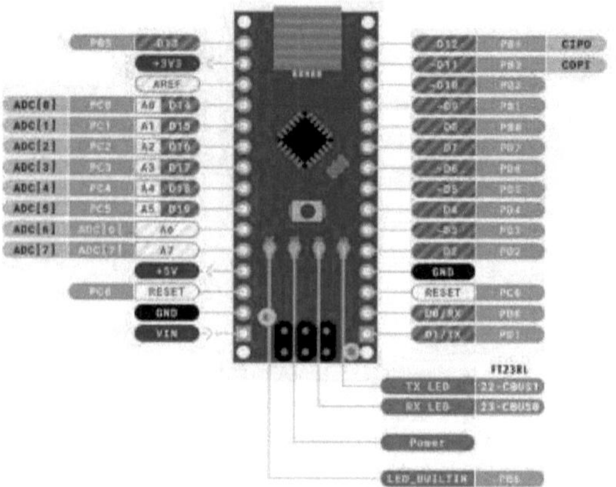

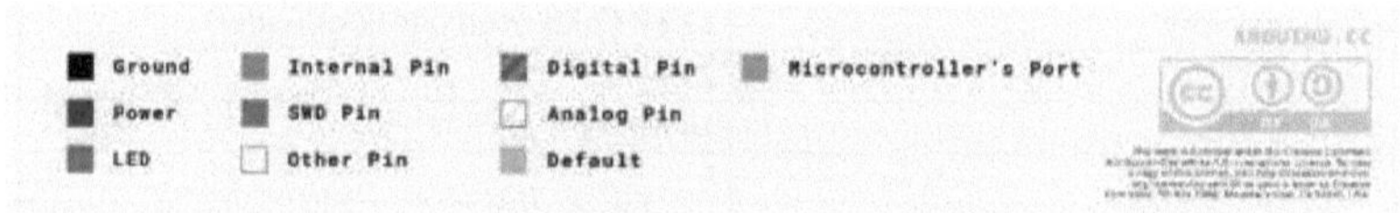

Figure N° 53Arduino Nano Pinout

2.2.2 Voltage regulator LM2596

The voltage regulator shown on page 42was selected because of its accuracy and its reduced cost and size. Adding this regulator to the PLC circuit allows to work with a wider voltage range, to operate with the most common voltages present in industrial controls (12/24VDC).

2.2.3 Relay output module

The relay output module has been selected (see page 49). 39) has been selected. Although the system requires only 4 actuator outputs, it is good design practice to consider additional outputs and inputs for future expansion of the automation.

2.2.3.1 Technical specifications of the 8-channel relay output module

Table N° 7. Technical specifications of 8-channel relay module

Category	Specification
Operating Voltage	5V DC
Control Signal	TTL (3.3V or 5V)
Number of relays (channels)	8 CH
Maximum capacity	10A/250VAC, 10A/30VDC
Maximum current	10A (NO), 5A (NC)
Time of action	10 ms (on) / 5 ms (off)
To activate output NO	0 Volts

2.2.4 Opto-coupled digital inputs

To electrically isolate the inputs from the rest of the PLC circuit, the PC817 integrated circuits, shown on page 53, were implemented. 43.

The design criteria for defining the number of inputs of our PLC has been the same as that considered for the relay outputs. Our system requires only 3 sensor inputs, but more inputs are left in reserve for future demands.

2.2.5 LCD display 2004

The display used is the LCD2004 model (Figure N° 54), based on the technology presented on page 50. 40. It can display 20 characters in 4 rows. It is one of the most economical options available for displaying data on a screen.

Figure N° 54LCD2004 screen

2.2.5.1 2004 LCD Technical Specifications

Table N° 8 8. Technical specifications of LCD 2004

Category	Specification
Module dimensions	98mm x 60mm
Character size	5mm x 8mm
Number of characters per line	20 characters per line
Number of lines	4 lines
Screen type	Alphanumeric character LCD
Operating voltage	5V
Connection interface	I2C serial interface
Backlight	LED
Contrast	Adjustable

2.2.6 DS3231 clock module

DS3231 is an extremely accurate and low-cost real time clock (RTC) with I2C communication and an integrated temperature compensated crystal oscillator.

The device incorporates a battery input, for when it is necessary to disconnect the main power supply and continue to maintain accurate timekeeping in the PLC. It maintains information of seconds, minutes, hours, day, date, month and year. In this way, the PLC is prepared to execute instructions that depend on a certain date or time.

The clock operates in 24-hour display or band/AM/PM 12-hour format. It provides two configurable alarms and a calendar that can be configured on a square wave output. Address and data are transferred serially via a bidirectional I2C bus.

A precision temperature-compensated voltage reference and comparator circuit monitors VCC status to detect power failures, provide a reset output and, if necessary, automatically switch to the backup power supply.

Figure N° 55Battery Module DS3231

2.2.6.1 Technical specifications of the DS3231 module

Table N° 9. Technical specifications of the DS3231 module

Category	Specification
Operating Voltage	3.3V - 5V
High-precision RTC	DS3231 with internal oscillator
Clock accuracy	2ppm
DS3132 I2C address	Read(11010001), Write(11010000)
EEPROM memory	AT24C32 (4K * 8bit = 32Kbit = 4KByte)
Communication	I2C
Square wave output	Programmable
Battery life	Can keep the RTC running for 10 years
Cascading compatibility	Can be used in cascade with another I2C device, the address of the AT24C32 can be changed (default is 0x57).

2.3 Circuit design

For the design of the PLC electronic circuits and the layout of the components on the PCB, the EasyEDA circuit design software was used. [28]which can be accessed free of charge and used online without the need to install any additional software.

It has the advantage of having a very complete library with the different electronic components and modules that can be obtained on the market, which greatly reduces design time. In addition, being online, it represents a cloud service in which files are protected and accessible from any device.

In the Figure N° 56 y Figure N° 57the two development environments offered by the software are shown. On the one hand, there is the environment for circuit design, where the components can be inserted from the library and the interconnections can be made.

Once the circuit is defined, you can use the "Convert schematic to PCB" function, which takes you to the PCB development environment. In it, you will find the traces of the selected electronic components, and interconnection guide lines, which show the existing links between components. From there, the PCB location and routing design is made, using the different tools and configurations.

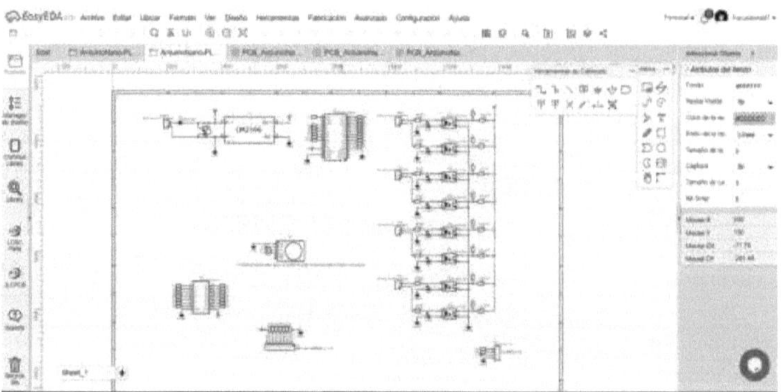

Figure N° 56EasyEDA circuit development environment

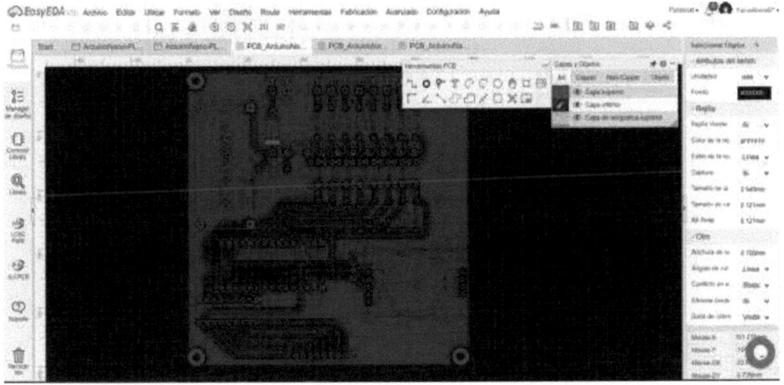

Figure N° 57PCB development environment EasyEDA

2.3.1 Arduino Nano Schematic

Based on the Arduino nano pinout diagram (Figure N° 53), the ports to be used were identified and the connection was made through network ports, which are flags that interconnect between those with the same name. This facilitates the design and reading of the electronic circuit.

The selection of the ports corresponding to each input and output were defined under an iterative design process, in which the directions of the connections had to be changed several times, in search of a more optimal design for the routing of the connections on the PCB.

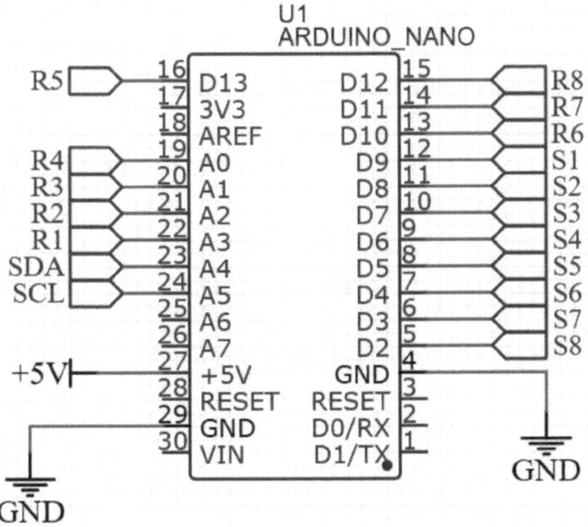

Figure N° 58Schematic diagram Arduino Nano

In the Figure N° 58we can identify the following connections:

Table N° 10. List of input connections

Pin Arduino	Puerto de red	Entrada
D9	S1	I1
D8	S2	I2
D7	S3	I3
D6	S4	I4
D5	S5	I5
D4	S6	I6
D3	S7	I7
D2	S8	I8

Table N° 11. List of output connections

Pin Arduino	Puerto de red	Salida
A3	R1	Q1
A2	R2	Q2
A1	R3	Q3
A0	R4	Q4
D13	R5	Q5
D10	R6	Q6
D11	R7	Q7
D12	R8	Q8

All signal pin connections on our Arduino Nano, both for outputs and inputs, are digital. In the Figure N° 58we can see that analog pins were used. This is possible because the analog pins of the Arduino Nano can be configured to be used as digital pins, with the exception of pins A6 and A7, which are exclusively analog.

2.3.2 Schematic diagram of opto-coupled inputs

The opto-coupled input circuit is a crucial step to allow the PLC to work with a wider variety of signal voltages, from 24V to 5V, being 24V and 12V the most common in the industrial automation field. In addition, it keeps the microcontroller protected from any overvoltage that may be generated on any of the input pins.

In the Figure N° 59 shows the schematic of the circuit corresponding to the digital inputs.

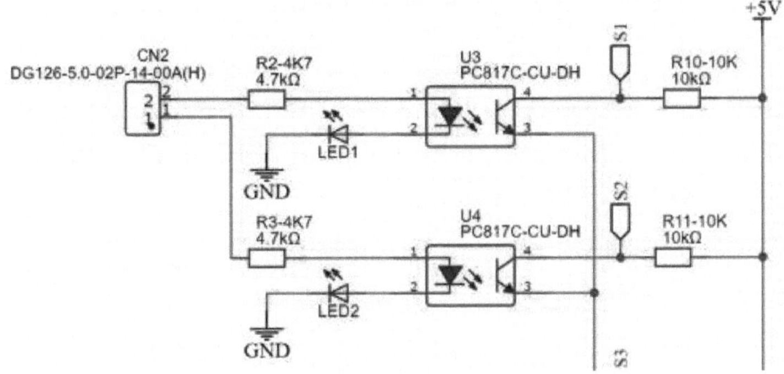

Figure N° 59Schematic diagram of opto-coupled inputs

In the circuit an input resistor of 4.7 Kohm can be identified, which regulates the input current to pin 1 of the optocoupler. At the output of pin 2, an LED is connected which indicates when an input signal is active. On the other side of the optocoupler, pins 3 and 4, there is a pull up configuration for switching the signal to the microcontroller, working at a voltage of 5V.

2.3.3 Schematic diagram of voltage regulator module

In Figure N° 60 shows the connections of the LM2596 module, which has been regulated by its integrated potentiometer to deliver a 5V output.

A rectifier diodo has been placed to protect the circuit in case the positive and negative terminals are connected in reverse. At the same time, a resistor in series with an LED was added in parallel to the input to indicate when the circuit is being energized correctly.

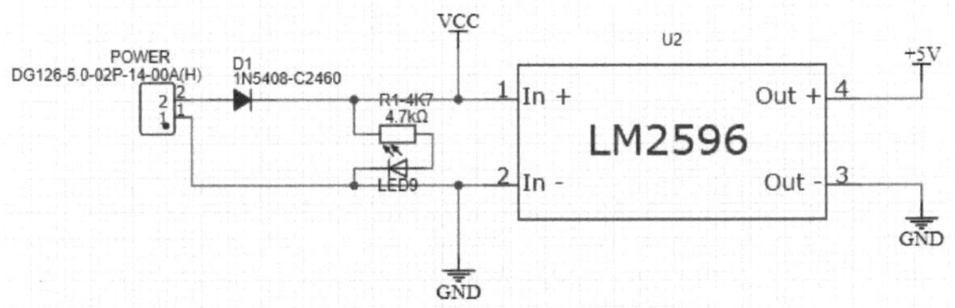

Figure N° 60Schematic diagram of the voltage regulator module

2.3.4 Schematic diagram of relay outputs

The Figure N° 61 shows the connection to the pins of the 8-channel relay output module. It can be seen that the input pins, coming from the Arduino Nano (R1 to R8), to the ULN2803 integrated unit, guarantee enough current to activate all the relays simultaneously if necessary.

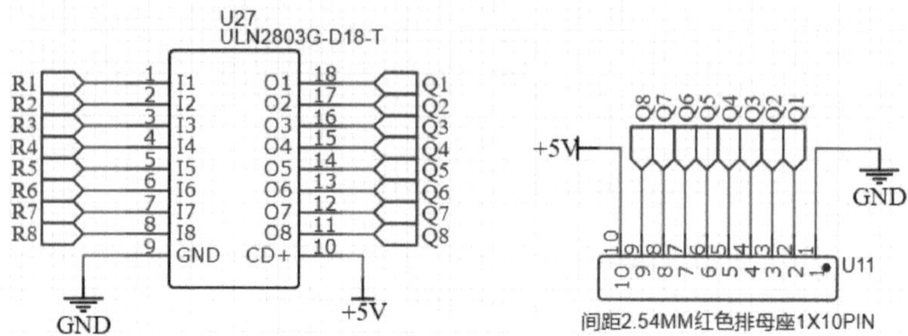

Figure N° 61Relay module connection

2.3.5 I2C port

The connections required for communication using the I2C protocol are shown in Figure 62. Figure N° 62. The SDA (Serial Data) and SCL (Serial Clock) signals from the Arduino nano were taken to a terminal block accessible to the devices to be communicated. In addition, two terminals with 5V and GND were added, so that the same reference voltage is shared in the communication.

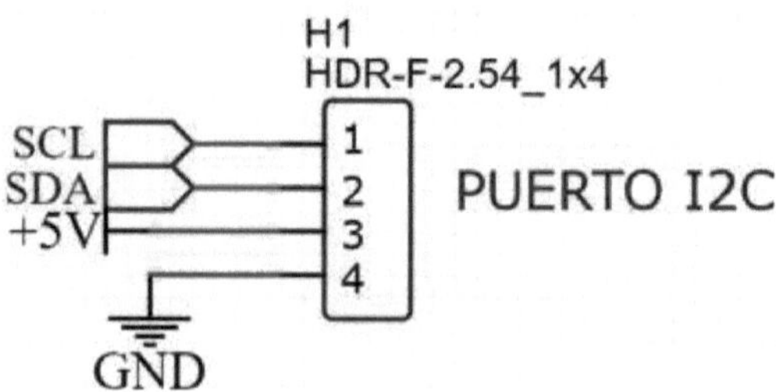

Figure N° 62I2C Port Connections

2.3.6 PCB design

Once the circuit schematic was ready, with its routing and electronic components, we proceeded to generate the printed circuit design. A board size of 100x100mm was selected, which adapts well to the size of the circuit and virgin copper boards of these dimensions are easily obtained in the market.

In the Figure N° 63the lower layer is shown in blue, corresponding to the copper area, and the upper layer in yellow, corresponding to the component silk-screen printing. In addition, some red connections have been added, which are bridges placed on the upper layer.

The main design criterion was to place the connection terminals on the top edge of the board, and the USB connector of the Arduino Nano board on the side edge, so that it could be easily accessed for programming. After this, the components were distributed in such a way as to optimize the use of the board area as much as possible.

A copper area has been added on the board, covering all the empty spaces, which is associated with the GND network. This is useful to capture electromagnetic noise, which can cause false signals in our circuit.

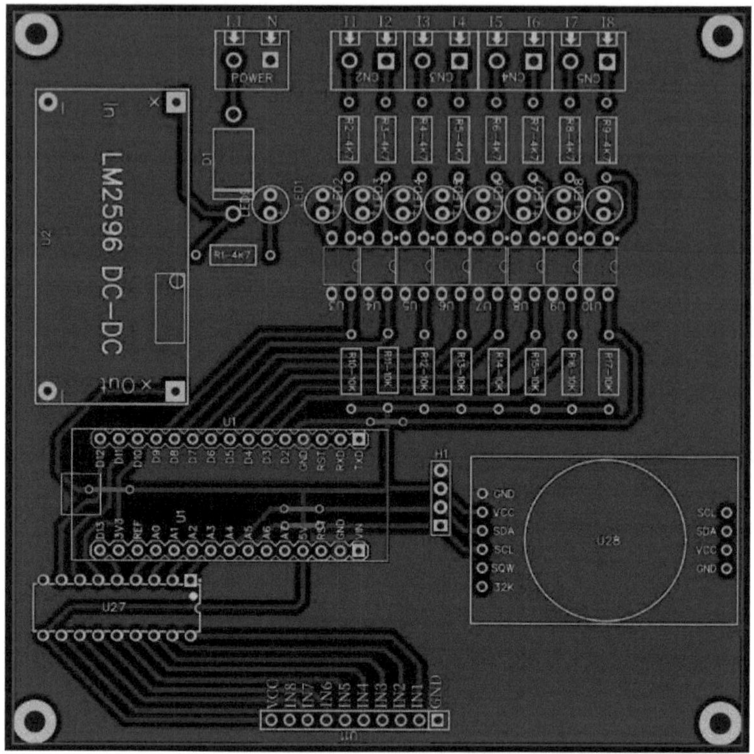

Figure N° 63PCB layout

The EasyEDA online software offers a 3D environment where a 3-D model of the PCB being designed is automatically generated (Figure N° 64 y Figure N° 65). This functionality is very useful for analyzing the positions of the electronic components and identifying any possible interference or design errors.

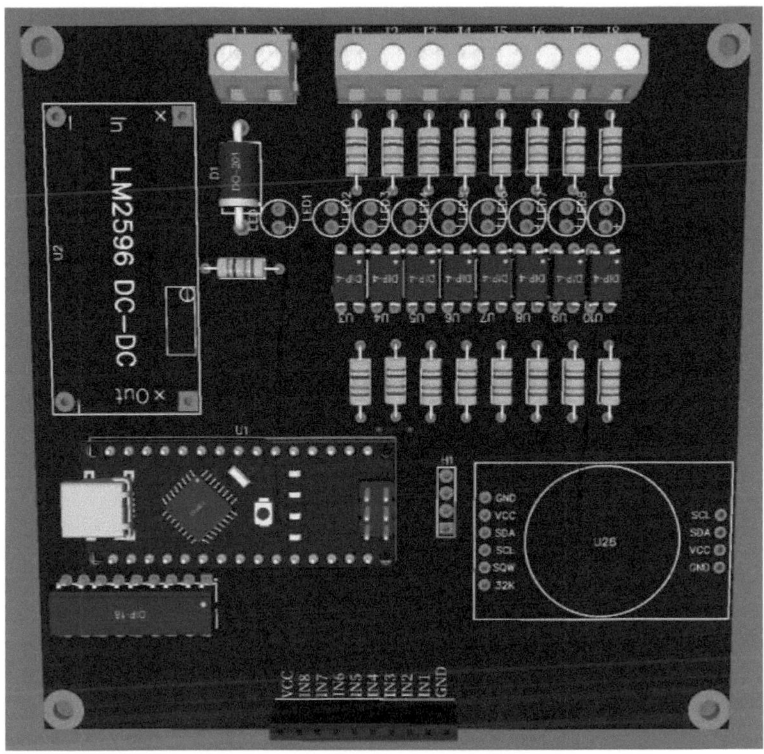

Figure N° 643D top view of PCB

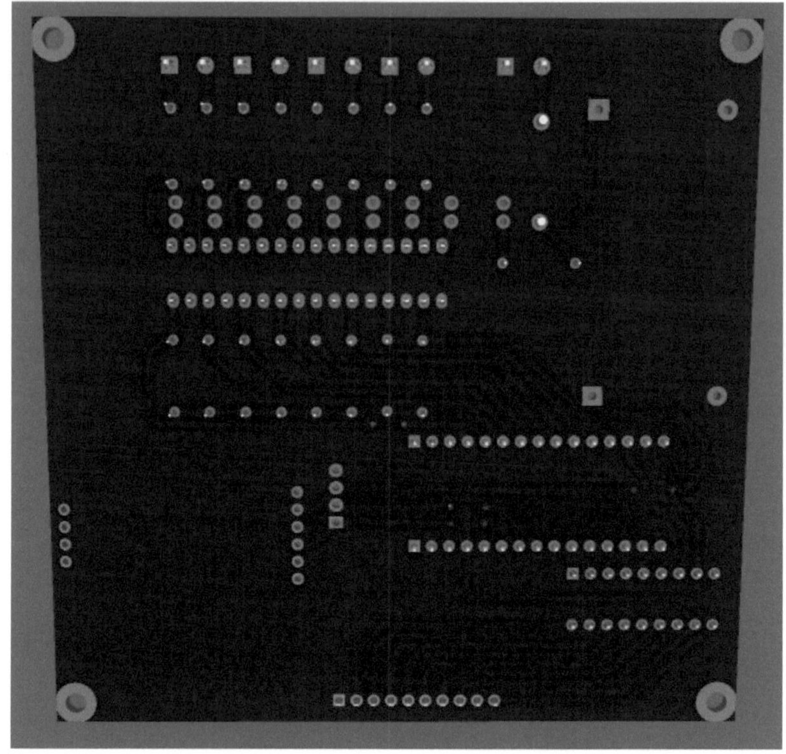

Figure N° 653D bottom view of PCB

2.3.6.1 Routing parameters

The routing rules were selected slightly larger than what is normally found in circuits of these powers, to ensure a good result during the board fabrication process by chemical etching.

The rules used are detailed below:

- Routing width: 0.7 mm
- Separation distance: 0.3 mm
- Track diameter: 1.2 mm
- Hole diameter: 0.8 mm

2.4 Electronic board manufacturing

2.4.1 PCB fabrication

Once the PCB circuit design was defined, it was fabricated using the chemical etching method. The following section will describe the steps carried out and the results obtained throughout the process.

2.4.1.1 Materials

- Virgin PCB board 100x100mm single sided Phenolic Pertinax
- Ferric perchloride acid
- Photographic paper
- Indelible ink marker
- Alcohol
- Virulana fine

2.4.1.2 First step: Thermotransfer engraving

The main key to a good result in PCB manufacturing is to ensure a correct thermal transfer. In this step, the top and bottom layers are printed on a sheet with a laser printer, since the ink in this printing adheres to the sheet by thermal transfer, from the heat produced by the laser. It is important that the sheet used is as smooth as possible, since rough sheets retain much more ink once they are printed. The ideal is to use thermotransfer paper, which is designed for this type of work, but since it was not available locally, photographic paper was used, with which a good result was obtained.

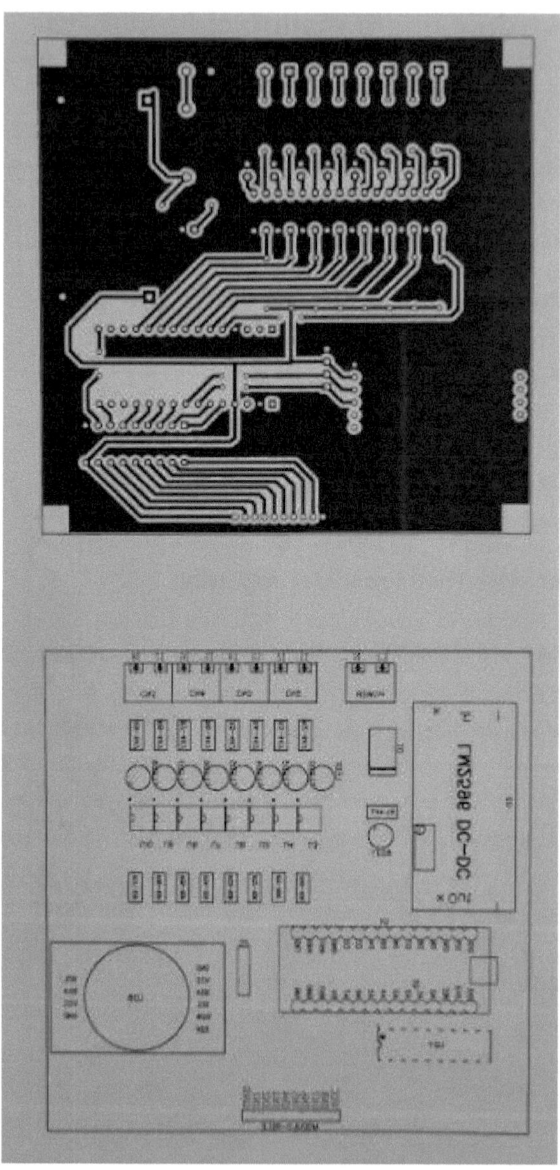

Figure N° 66PCB printing on photographic paper

With the printing of the lower layer, placing the ink side on the copper side of the virgin plate, the thermotransfer was generated. The heat was supplied by constant pressure with an iron for 10 minutes.

After this, with the ink already transferred from the paper to the copper side, the plate is immersed in water to undo the paper and remove all its remains. The result is shown in Figure N° 67.

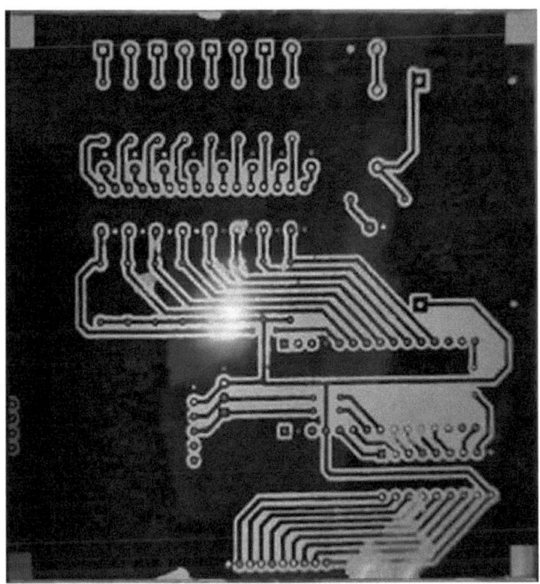

Figure N° 67Dye transfer to blank plate

It is possiblo to observe certain areas where the ink transfer was not complete. To correct these details, an indelible black marker was used.

2.4.1.3 Second step: Acid immersion

With the PCB design already transferred, the copper plate is immersed in ferric perchloride. This acid attacks the copper that comes in contact with it, dissolving it. Thus, all the area that is not painted will be swept away, leaving only the copper that is protected by the ink.

This process requires about 10 minutes of exposure to the acid. Moving the plate continuously speeds up the process, and preheating the acid to about 45 degrees allows it to be more reactive.

Once the acid has acted, the plate is removed and rinsed with water. To remove the ink residues, a fine virulana and a cloth with alcohol are used. The final result is shown in Figure N° 68.

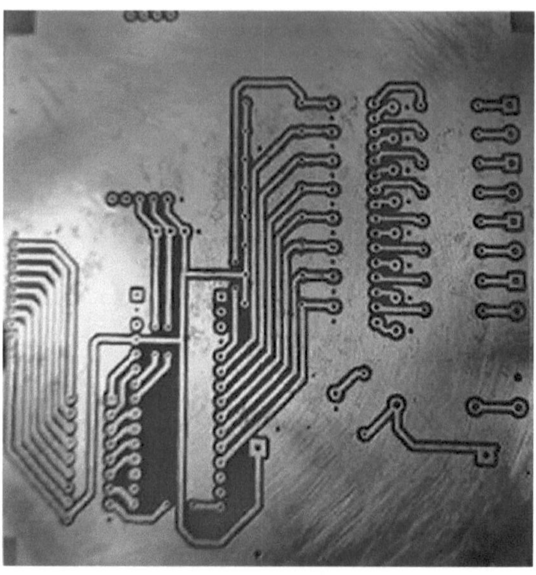

Figure N° 68Finished PCB board

2.4.1.4 Third step: Screen printing

With the same thermotransfer method used for the lower side, the silkscreen is transferred to the upper side of the plate, which is the one without copper.

Screen printing is of great importance to know the location of the electronic components during the soldering process, and to provide important circuit information.

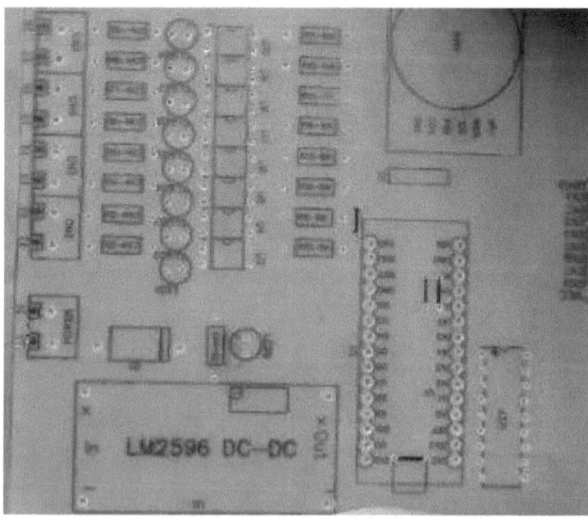

Figure N° 69PCB silkscreen printing

2.4.2 Assembly of electronic components

Once the PCB board was ready, it was drilled with a 0.3 mm drill bit. All the electronic components, which are THT (Through-Hole Technology), were mounted on the top layer, using the silk-screen printing as a guide, and soldered on the copper side, on the bottom layer.

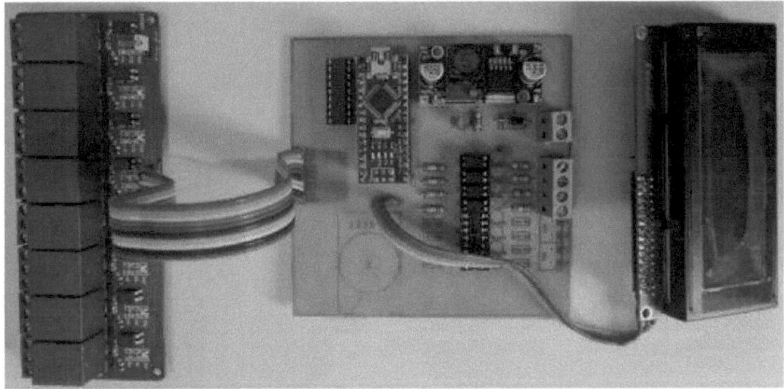

Figure N° 70Final electronic board

2.5 3D modeling of casing

The 3D parametric modeling software "Inventor" from Autodesk was used to design the housing.

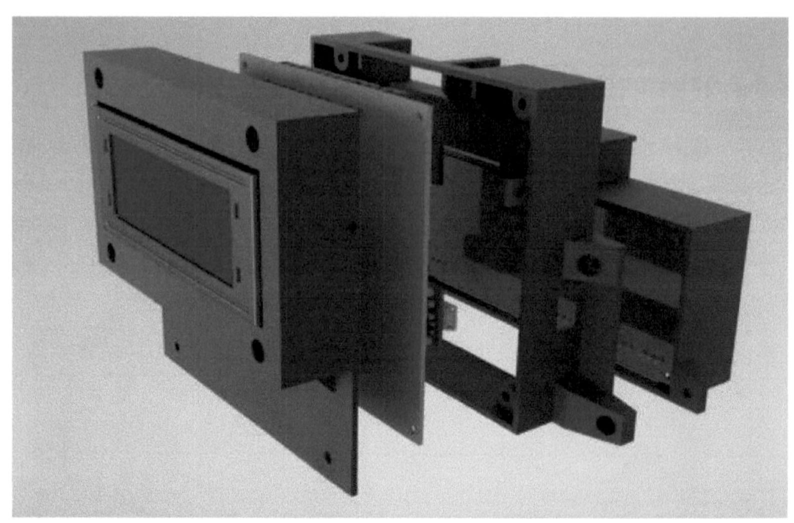

Figure N° 713D model of casing

The housing design is composed of 3 parts, one of which houses the relay module, another central part where the electronic board is located, and a front part which is where the LCD display is mounted.

Each of them are joined together by means of M3 bolts and nuts (Figure N° 72).

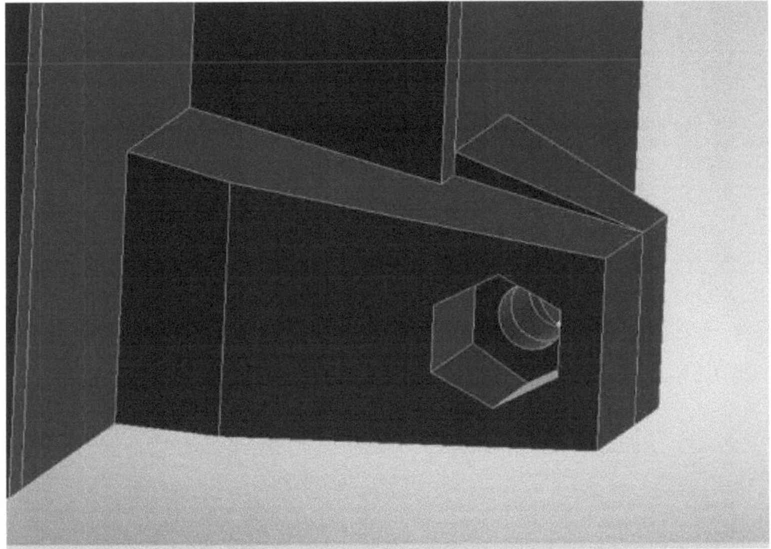

Figure N° 72: Housing for M3 nut

By means of 3D printing, it was possible to materialize the parts with an acceptable finish. This manufacturing method provides flexibility when designing, as long as the difficulties or limitations involved are taken into account.

PLA filament was used as material, which presents a good balance between hardness and flexibility, and does not show great difficulties during the printing process. The wall thickness is 2mm and the filling pattern is honeycomb.

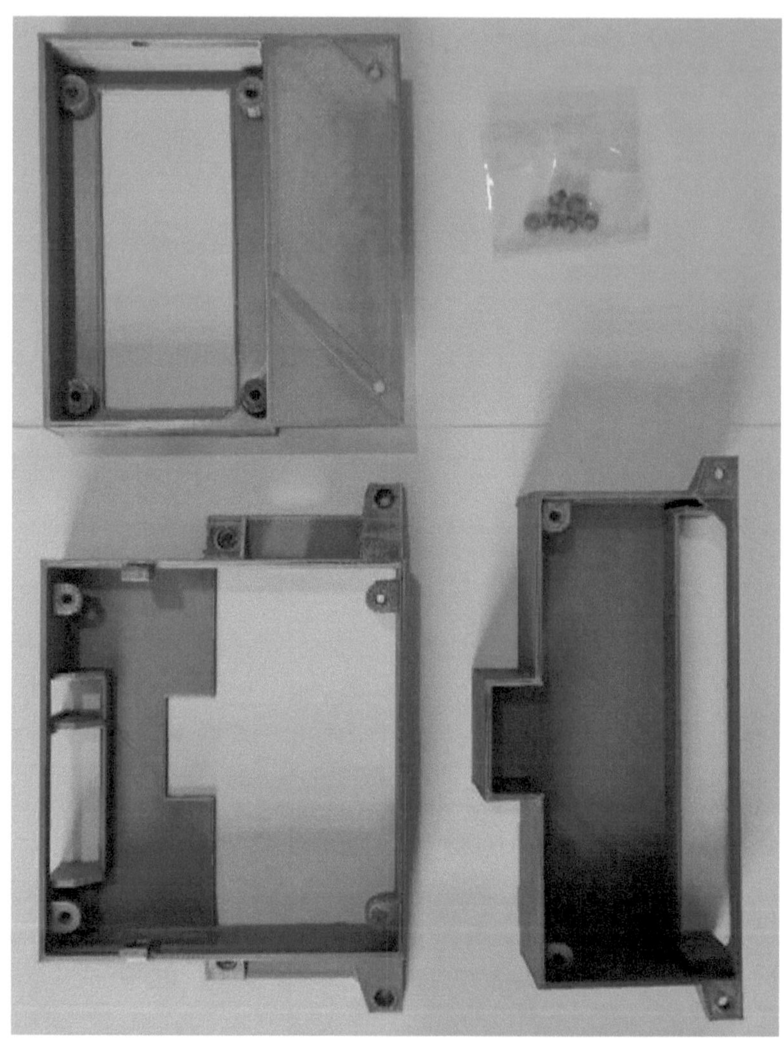

Figure N° 73PLC housing parts

Figure N° 74PLC Assembly - Front View

Figure N° 75PLC Assembly - Top view

Figure N° 76PLC Assembly - Rear View

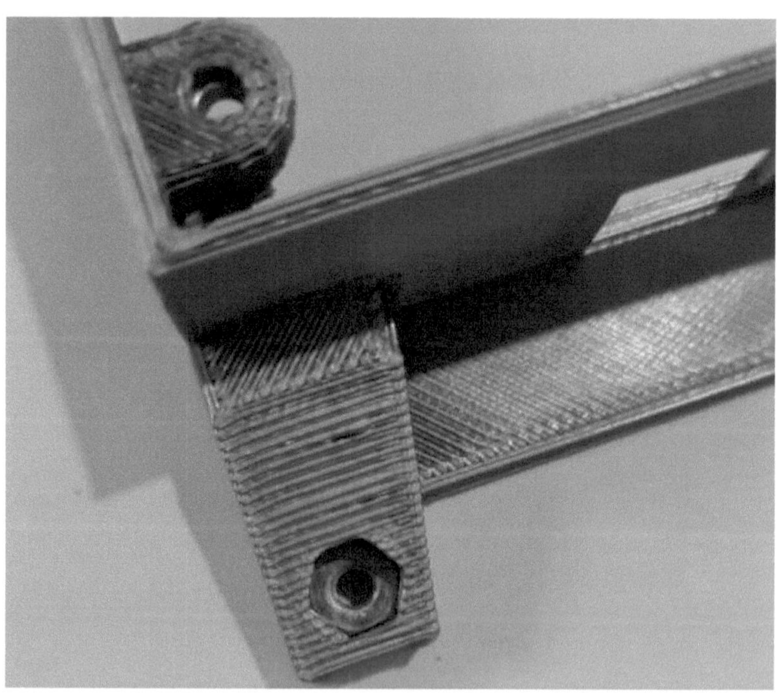

Figure N° 77Nut insert detail

2.5.1 Programming logic

In this section, all aspects related to the programming involved in the project will be described.

Before moving on to the lines of code, a state machine had to be created to precisely define the logic and operation of our automation system.

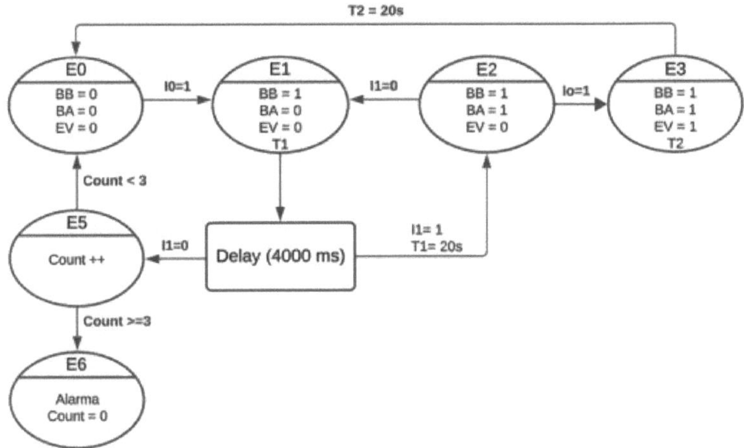

Figure N° 78State machine of the automation system

From this diagram (Figure N° 78), progress was made on the coding in C++ language.

2.5.1.1 State machine analysis

E0:

- Description: This is the initial state or idle state, in which all outputs are disabled.
- Transitions: When the float (I0) indicates low level in the buffer tank, it transitions to the E1 state to start the process.

E1:

- Description: The low pressure pump (BB) is switched on, to reach the suction pressure of the high pressure pump (BA). Timer T1 is started.
- Transitions: After 4 seconds of waiting, it is evaluated whether the pressure switch (I1) reached the proper pressure. This delay is necessary to avoid false readings caused by turbulence at start-up. If the pressure is correct (I1=1), and the time T1 = 20 s has been reached, it goes to state E3. If the pressure is incorrect (I1=0), it goes to state E5 to try a new start.

E2:

- Description: This is the "Equipment Running" state, in which the high pressure pump is turned on in conjunction with the low pressure pump.
- Transitions: If the pressure is low (I1=0), it goes back to state E1. If the float indicates that the tank level is high (I0=0), it goes to state E3.

E3:

- Description: This is the "Backwash" state, in which the solenoid valve is opened and the high and low pressure pump is kept on. The T2 timer is started.
- Transitions: When T2= 20 s is satisfied, it goes to the E0 state to turn off all outputs.

E5:

- Description: This status keeps track of failed startup attempts (Count).
- Transitions: If Count < 3, it goes to the initial state E0. If Count >=3, it goes to state E6.

E6:

- Description: This is the alarm status. It indicates that there is a problem in the system and the equipment has failed to start after 3 attempts.
- Transitions: It does not have any transition. This state requires system intervention and revision. The equipment must be restarted to exit the alarm state.

2.5.2 Programming code

Below is the programming code written in Arduino IDE, which has been loaded into the Arduino Nano microcontroller to execute the control of the water treatment equipment.

```
#include <Arduino.h>
#include <Wire.h>
#include <LiquidCrystal_I2C.h>

LiquidCrystal_I2C lcd(0x27, 20, 4); // I2C address 0x27, 20 column and 4
rows

// Pines PLC Arduino NANO
// ENTRADAS(Pull up):
#define I0 9
#define I1 8
#define I2 7
#define I3 6
#define I4 5
#define I5 4
#define I6 3
#define I7 2

// Salidas
#define Q0 A7
#define Q1 A6
#define Q2 A3
#define Q3 A2
#define Q4 A1
#define Q5 A0
#define Q6 13
#define Q7 12

const int flotante = I0;  // Flotante de baja de tanque de agua tratada --
0v: tanque lleno. 24v: pide agua
const int presostato = I1;
const int rele1 = Q7;                // Bomba Baja
const int rele2 = Q6;                // Bomba Alta
const int rele3 = Q5;                // Electrovalvula
const int rele4 = Q4;                // Alarma1

unsigned long T1 = 0;
```

```
unsigned long T2 = 0;
unsigned long T3 = 0;
unsigned long periodo2 = 20000;  //Periodo de retrolavado
int count = 0;                   //Contador de intentos de arranque fallidos
int Estado = 0;
int Sig_Estado = 0;
int screen = 0;                 //Variable que maneja las diferentes
pantallas(textos) a mostrar

void setup() {
  lcd.init(); // initialize the lcd
  lcd.backlight();
  pinMode(flotante, INPUT);
  pinMode(presostato, INPUT);

  //Inicio el programa con todo apagado
  digitalWrite(rele1, LOW);  // Apaga Bomba Baja
  digitalWrite(rele2, LOW);  // Apaga Bomba Alta
  digitalWrite(rele3, LOW);  // Cierra Electrovalvula
  digitalWrite(rele4, LOW);  // Apaga Alarma1: Intento fallido de arranque

  pinMode(rele1, OUTPUT);
  pinMode(rele2, OUTPUT);
  pinMode(rele3, OUTPUT);
  pinMode(rele4, OUTPUT);

}

void loop() {



  switch (Estado) {
    case 0: // Estado inicial
      digitalWrite(rele1, LOW);  // Apaga BB
      digitalWrite(rele2, LOW);  // Apaga BA
      digitalWrite(rele3, LOW);  // Cierra Electrovalvula
      digitalWrite(rele4, LOW);  // Apaga Alarma1: Intento fallido de
arranque
      if (digitalRead(flotante) == HIGH){
      lcd.setCursor(0, 0);
      lcd.print("EQUIPO APAGADO");
      lcd.setCursor(0, 2);
      lcd.print("->Tanque lleno");
      delay(3000);
```

```
      Sig_Estado = 1;
        T1 = millis(); //Se inicia el temporizador 1
        delay(3000);
        lcd.clear();
      }
      break;

    case 1: // Estado de encendido de la bomba baja
      digitalWrite(rele1, HIGH);
      digitalWrite(rele2, LOW); //Bomba de alta apagada
      lcd.setCursor(0,0);
      lcd.print("ARRANCANDO EQUIPO..");
      delay(4000);  //Delay para evitar falsas lecturas del presostato
      if (digitalRead(presostato) == LOW) {  //Si la presion subio luego del
delay, se continua con el proceso de encendido
        if (millis() > T1 + periodo) {  // ¿por que hay que esperar tanto
tiempo para que la BA arranque?
          Sig_Estado = 2;
          count = 0;
          lcd.clear();
        }
      } else if (digitalRead(presostato) == HIGH) {  //Si el presostato
indica baja presion luego de haber encendido la BB, se apaga BB, se
incrementa el contador y se vuelve al estado 0
        digitalWrite(rele1, LOW);  //Se apaga BB
        count++;
        if (count >= 3) { //Se superaron los 3 intentos fallidos de arranque
          Sig_Estado = 6; // Estado de alarma
          lcd.clear();
        } else if (count < 3) {
          Sig_Estado = 0; // Volver al estado inicial para un nuevo intento
          lcd.clear();
        }
      }
      break;

    case 2: // Estado de encendido de la bomba alta
      digitalWrite(rele2, HIGH);
      lcd.setCursor(0, 0);
      lcd.print("EQUIPO FUNCIONANDO");
      lcd.setCursor(0, 2);
      lcd.print("->Llenando tanque");
      if (digitalRead(presostato) == HIGH) { //Si hay algun problema con la
presion, se regresa al estado 1 para volver a intentar el arranque
        Sig_Estado = 1;
```

```
        }
      if (digitalRead(flotante) == HIGH) { //El flotante me indica detener
el bombeo
        Sig_Estado = 3; // Estado de apagado
        T2 = millis();  //Inicio contador T2
        lcd.clear();
      }
      break;

    case 3: // Estado de apertura de electrovalvula
      digitalWrite(rele3, HIGH); //Abre Electrovalvula
      digitalWrite(rele2, LOW); //Apaga BA
      lcd.setCursor(0, 0);
      lcd.print("AUTOLIMPIANDO..");
      if (millis() > T2 + periodo2) {
        Sig_Estado = 0;
        T3 = millis();
        lcd.clear();
      }
      break;

    case 6: // Estado de alarma
      digitalWrite(rele4, HIGH);
      lcd.setCursor(0, 0);
      lcd.print("ALARMA!");
      lcd.setCursor(0, 2);
      lcd.print("->Arranque fallido");
      // Este estado se mantiene hasta que se reinicie el Arduino
      break;
  }

  Estado = Sig_Estado;
}
```

The I2C protocol was used to connect the PLC with the LCD display, which has its own library that facilitates its programming and configuration:

```
#include <LiquidCrystal_I2C.h>

LiquidCrystal_I2C lcd(0x27, 20, 4); // I2C address 0x27, 20 column and 4
rows
```

The screen was configured based on the size (20 columns by 4 rows) and assigned the address 0x27, which is in hexadecimal format and corresponds to the address of the screen, provided by the manufacturer.

The pins corresponding to the inputs and outputs were defined, based on the hardware configuration chosen in the PCB design:

```
// Pines PLC Arduino NANO
// ENTRADAS(Pull up):
#define I0 9
#define I1 8
#define I2 7
#define I3 6
#define I4 5
#define I5 4
#define I6 3
#define I7 2

// Salidas
#define Q0 A7
#define Q1 A6
#define Q2 A3
#define Q3 A2
#define Q4 A1
#define Q5 A0
#define Q6 13
#define Q7 12
```

2.6 Development of industrial communication module

A crucial part in the development of this automation solution is the ability to manage process data. For this purpose, an industrial communication module was developed to link the PLC with any other device connected to the Internet.

In this section, the development of the module will be described, together with all the agents involved in the communication system.

Because the communication module will not be implemented in conjunction with the PLC that controls the water purification equipment, communication with the PLC has been simulated using the Arduino Nano microcontroller.

2.6.1 System architecture

The Figure N° 79 shows the components involved in industrial communication and the protocols that were implemented.

The PLC developed has an I2C communication port available on terminal boards, so the communication between the Arduino Nano microcontroller and the communication module will be through this protocol, in case that in the future it is desired to incorporate data monitoring. The variables monitored in this case are water flow and conductivity.

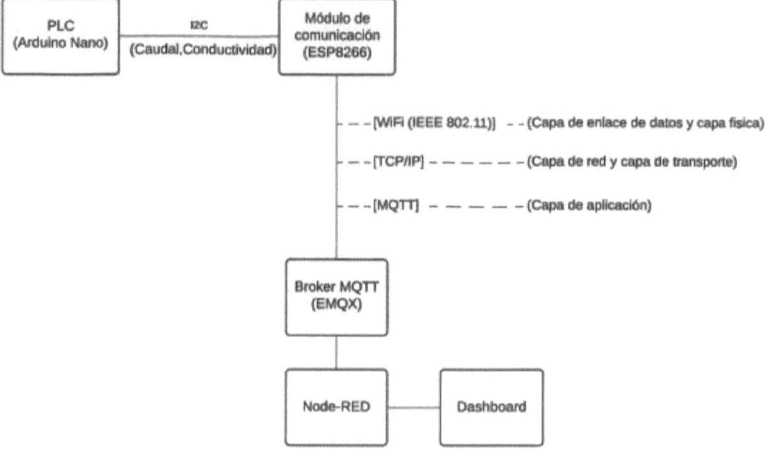

Figure N° 79 Communication system architecture

Each of the elements and their linkages will be discussed below.

2.6.1.1 Communication module

The communication module is in charge of interfacing the PLC with any other device available on the Internet. It takes the flow and conductivity data transmitted by the PLC through the I2C protocol, and retransmits them to the MQTT broker using the following protocols:

- WiFi (IEEE 802.11): Provides network connectivity, in this case wirelessly, between the device and the router or access point.
- TCP/IP: Provides the fundamental network communication for communication over the Internet and IP networks.
- MQTT: Provides lightweight and efficient messaging between devices and applications.

To manage this interconnection, the ESP8266 microcontroller was used, which has very similar characteristics to the Arduino Nano, and in turn integrates WiFi and Bluetooth connectivity capabilities. It can be programmed in C++ using the Arduino IDE.

Figure N° 80ESP8266 Microcontroller

2.6.1.2 Broker MQTT

The broker acts as an intermediary between the devices that publish messages (ESP8266) and the devices that subscribe to those messages (Node-RED). It is responsible for receiving all messages, filtering them and redirecting them to interested subscribers.

The broker can be installed on a computer that is connected to the same network as the publishers and subscribers, or it can be used, as in this case, a broker in the cloud that can be accessed while connected to the Internet, without the need to install any additional software. The one used in this project is EMQX. [29]

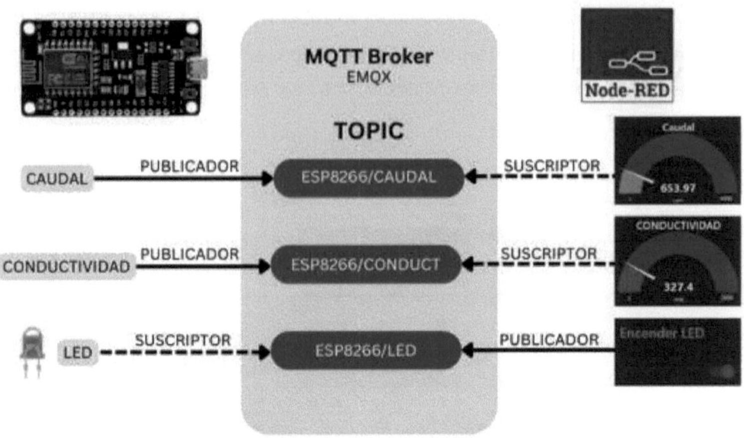

Figure N° 81MQTT connection diagram

The Figure N° 81 shows how the information linkage between the ESP8266 and Node-RED is performed through the EMQX broker. It can be seen how the subscriber/publisher method works, where each publisher updates the value of a given topic, and then each subscriber takes the value of the topic. There is no direct connection between publisher and subscriber, which allows the same data to be taken by different devices, and at the same time different devices can publish in the same topic.

To visualize the bidirectional communication between the ESP8266 and Node-RED, a switch was added that changes the value between 1 and 0 of the "ESP8266/LED" topic, and the ESP8266 takes that value to turn on or turn off the integrated LED.

2.6.1.3 Node-RED

Node-RED is the development tool used to visualize the data of our process in a Dashboard. In the Figure N° 82 shows the flow developed for our purpose, where we can highlight the nodes corresponding to the MQTT connection for both subscriber and publisher, and the data visualization functions. A switch for activating and deactivating the ESP8266 LED has also been included here.

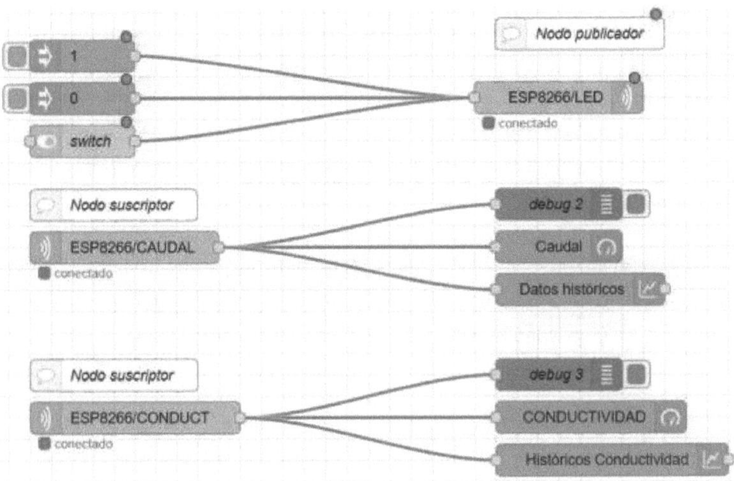

Figure N° 82Data flow in Node-RED

2.6.1.4 Dashboard

The Figure N° 83 shows the dashboard developed, where the flow and conductivity data are displayed by implementing a dial type graph, which shows the current value of the variable, and a line graph where the historical data can be reviewed and the changes produced can be analyzed. There is also a switch that turns the LED on or off.

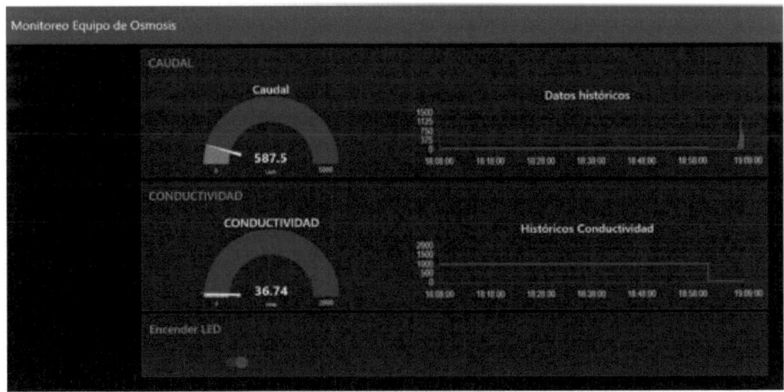

Figure N° 83Dashbboard of osmosis equipment

2.6.2 Prototype communication module

The Figure N° 84 shows the prototype conceived with the objective of verifying the correct operation of all the elements that make up the Communication system architecture (Figure N° 79). For this test, the data used were the flow and conductivity variables.

The electrical connections implemented are described in the schematic in Figure N° 85. Figure N° 85. The SDA and SCL paths that allow I2C communication between the Arduino Nano and the ESP8266 can be observed there.

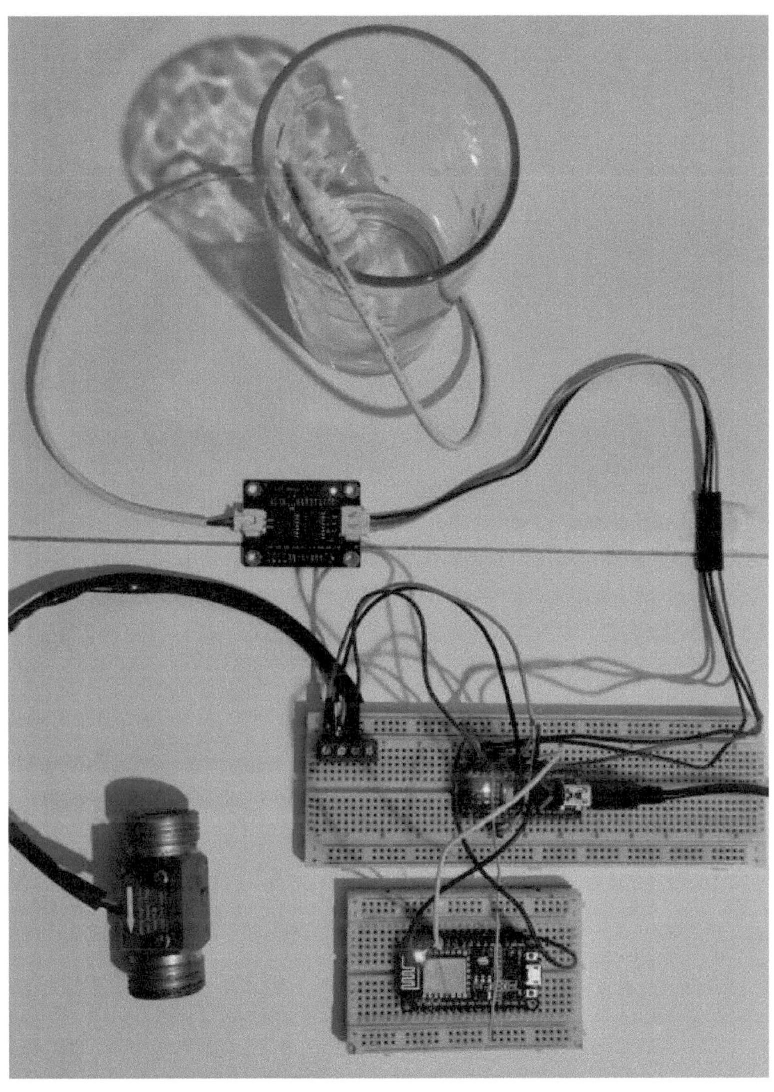

Figure N° 84Communication module prototype

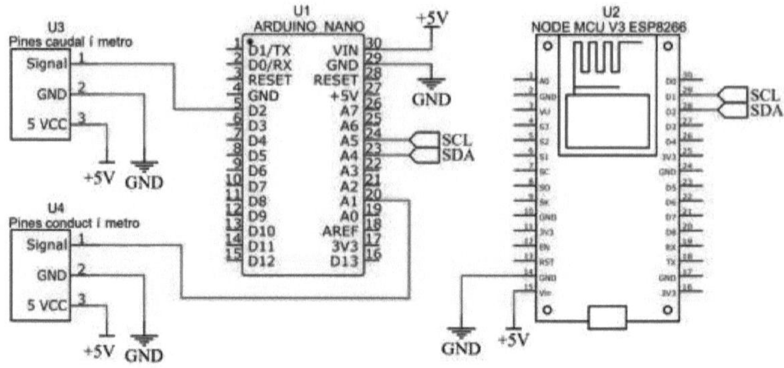

Figure N° 85Schematic - Communication module prototype

The connection was effective in all stages. The values sensed by the flowmeter and conductivity meter were correctly read, the variables were passed from the Arduino Nano to the ESP8266 by I2C protocol, and through WiFi, using the MQTT protocol, the data was transferred to the Node-RED to be displayed on the dashboard of the ESP8266. Figure N° 83. In addition, the correct operation of the switch to turn on the ESP8266 integrated led was tested, resulting in a solid bidirectional communication of the communication module.

This configuration was used as a test for subsequent implementation in a specific real development, ensuring the feasibility and effectiveness of the system in a specific industrial environment.

CAPITULO 3: COST ANALYSIS

In this chapter, the costs involved in the development and implementation of the PLC and the industrial communication module will be analyzed. The data will be contrasted with other devices and services available in the market, in order to establish a profitability margin.

The focus of this project is on the economic viability and democratization of industrial automation, so the key to the device being able to enter the market in future projects and represent an automation solution lies in the final cost of the product being lower than competing options.

The values expressed in dollars (USD) have been taken in reference to the value of the official dollar in Argentina, corresponding to 929.5 ARS on 06/26/2024.

3.1 PLC Costs

3.1.1 Direct costs

3.1.1.1 Hardware components

Table N° 12PLC Costs - Hardware Components

Item	Descripción	QTY	ARS	USD	ARS2	USD2
Carcasa	Impresión 3D	1	$ 16.000,00	USD 17,21	$ 16.000,00	USD 17,21
Pantalla	LCD 2004 c/adapt I2C	1	$ 13.570,00	USD 14,60	$ 13.570,00	USD 14,60
Modulo Rele	8 canales	1	$ 12.580,00	USD 13,53	$ 12.580,00	USD 13,53
Arduino nano	Microcontrolador	1	$ 7.019,00	USD 7,55	$ 7.019,00	USD 7,55
Placa de cobre	100x100mm	1	$ 3.500,00	USD 3,77	$ 3.500,00	USD 3,77
ULN2803	Puente Darlington x8	1	$ 3.420,00	USD 3,68	$ 3.420,00	USD 3,68
PC817	Optoacoplador	8	$ 401,50	USD 0,43	$ 3.212,00	USD 3,46
Bornera	2 pines	5	$ 530,40	USD 0,57	$ 2.652,00	USD 2,85
LM2596	Regulador 5V	1	$ 2.360,00	USD 2,54	$ 2.360,00	USD 2,54
Tornillos	M3 x 10mm	14	$ 104,00	USD 0,11	$ 1.456,00	USD 1,57
Tuercas	M3	14	$ 65,00	USD 0,07	$ 910,00	USD 0,98
LED	Verde rectangular	9	$ 93,60	USD 0,10	$ 842,40	USD 0,91
Zocalo 8 pines	Para PC817	4	$ 130,00	USD 0,14	$ 520,00	USD 0,56
R4k7	Resistencia 4,7kohm	9	$ 57,00	USD 0,06	$ 513,00	USD 0,55
R10k	Resistencia 10kohm	8	$ 30,00	USD 0,03	$ 240,00	USD 0,26
Zocalo 2x9 pines	Para ULN2803	1	$ 227,50	USD 0,24	$ 227,50	USD 0,24
1N4007	Diodo rectificador	1	$ 49,40	USD 0,05	$ 49,40	USD 0,05
				Total	$ 69.071,30	USD 74,31

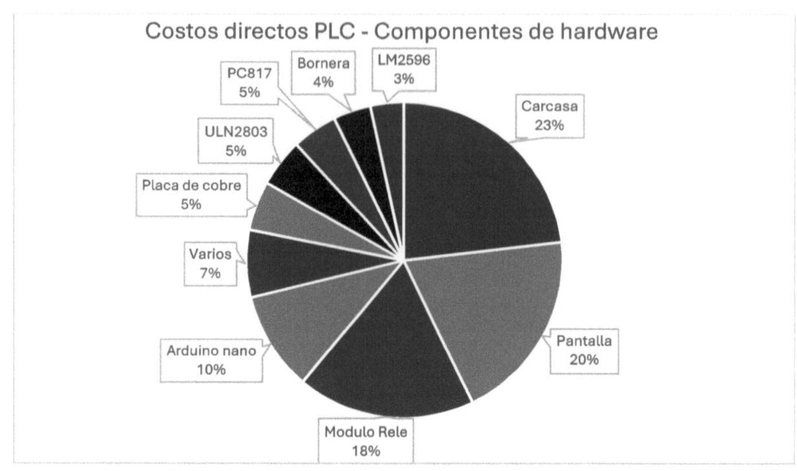

Figure N° 86Percentage of direct PLC costs - hardware components

3.1.1.2 Manufacturing costs

Manufacturing costs correspond to the labor, time and materials involved in assembling the electronic components to manufacture the PCB, and in assembling the final PLC assembly.

Table N° 13PLC manufacturing costs

			Costo unitario		Costo total	
Item	**Descripcion**	**QTY**	**ARS**	**USD**	**ARS**	**USD**
Ensamblaje	Hs de mano de obra	3	$ 4.000,00	USD 4,30	$ 12.000,00	USD 12,91
Estaño	Para soldadura	0,2	$ 4.550,00	USD 4,90	$ 910,00	USD 0,98
Acido	Percloruro ferrico	0,15	$ 9.100,00	USD 9,79	$ 1.365,00	USD 1,47
				Total	$ 14.275,00	USD 15,36

3.1.1.3 Software costs

As can be seen in Table Table N° 14The software used in the project is totally free, which provides a great competitive advantage.

Table N° 14 : Software costs

Item	**Description**	**Cost**
Arduino IDE	Microcontroller programming	Free
EasyEDA	Printed circuit board design	Free
Inventor	3D housing design	Free student license

Thus, taking into account hardware component costs, manufacturing costs and software costs, we obtain a total direct cost of **$83,346.30 (USD 89.67).**

3.1.2 Indirect costs

Indirect costs include the time and resources dedicated to the research and development of the project.

The calculations are based on the number of hours spent in each area at a value of $5000 ARS per hour.

Table N° 15Indirect costs - Research and Development

			Costo unitario		Costo total	
Item	**Descripcion**	**QTY (Hs)**	**ARS**	**USD**	**ARS**	**USD**
I+D	Investigacion y Desarrollo	72	$ 5.000,00	USD 5,38	$ 360.000,00	USD 387,31
PCB	Diseño de PCB	42	$ 5.000,00	USD 5,38	$ 210.000,00	USD 225,93
Programación	Programa de PLC en C++	36	$ 5.000,00	USD 5,38	$ 180.000,00	USD 193,65
Diseño 3D	Modelado 3D de carcazas	48	$ 5.000,00	USD 5,38	$ 240.000,00	USD 258,20
Compras	Busqueda de proveedores	18	$ 5.000,00	USD 5,38	$ 90.000,00	USD 96,83
				Total	$ 1.080.000,00	USD 1.161,9

3.2 Communication module costs

Since the development of the communication module only contemplates achieving effective communication between devices using the MQTT protocol and Ethernet, and not the implementation in an industrial environment, the scope of the cost analysis for this case will be limited to the necessary hardware components and the software involved.

3.2.1 Hardware components

Within the costs of hardware components of the communication module, the elements required to assemble the device inside a container or housing that allows mounting in an industrial environment are estimated.

Table N° 16Hardware components - Communication module

Item	Descripción	QTY	Costo unitario				Costo total			
			ARS		USD		ARS2		USD2	
ESP8266	Microcontrolador	1	$	6.090,00	USD	6,55	$	6.090,00	USD	6,55
LM2596	Regulador 5V	1	$	2.360,00	USD	2,54	$	2.360,00	USD	2,54
Carcaza	Impresión 3D	1	$	6.000,00	USD	6,46	$	6.000,00	USD	6,46
Bornera	2 pines	2	$	530,40	USD	0,57	$	1.060,80	USD	1,14
Placa cobre	50x50mm	1	$	3.500,00	USD	3,77	$	3.500,00	USD	3,77

Total	$ 19.010,80	USD 20,45

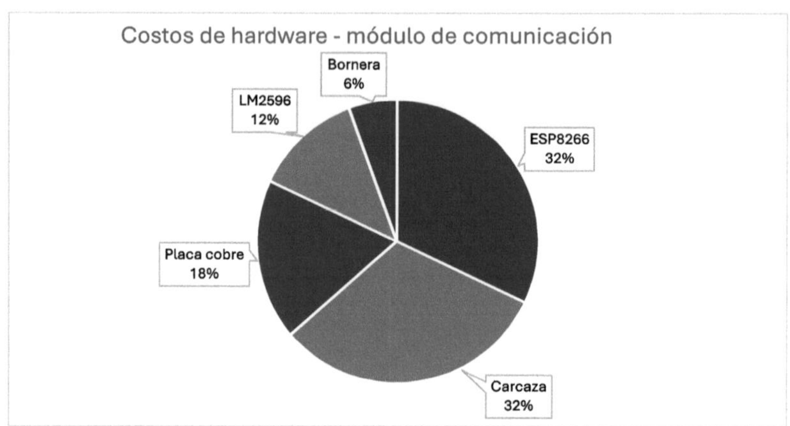

Figure N° 87Hardware costs - communication module

3.2.2 Software costs

Table N° 17Software costs - communication module

Item	Description	Cost
Arduino IDE	Microcontroller programming	Free
Inventor	3D housing design	Free student license
Node-RED	Connected device gestures	Free
EMQX	MQTT Broker in the cloud	Free

The software used in the development of the communication module is free and open source. However, many of the interesting functionalities for interconnected projects, such as data visualization and cloud storage, require further development or the incorporation of paid services.

CAPITULO 4: CONCLUSIONS

4.1 Summary of results

The main objective of this project has been the development of a PLC based on open technology and its implementation in an industrial environment, aiming to provide an effective and accessible automation solution. It was achieved that it complies with the established requirements, and that it works in permanent regime in the control of the reverse osmosis equipment.

On the other hand, the focus of this project has been for the device to communicate with other systems, allowing data management and the creation of automation networks. This was achieved with the prototype of the industrial communication module, which demonstrated excellent performance in the control and monitoring of the device remotely, with linkage through the Internet.

4.2 Evaluation of compliance with objectives

4.2.1 General Objective

1. Develop and implement a low-cost Programmable Logic Controller (PLC) using open technologies.

    - **Fulfillment:** An Arduino based PLC has been developed and implemented, using open components and technologies. The total cost of the device has been significantly lower compared to commercial solutions, thus meeting the low cost objective.

2. To use the device in an industrial environment for the effective control of a water treatment equipment, with a permanent operating regime.

    - **Compliance:** The PLC has been installed in an industrial environment for the control of the reverse osmosis equipment, operating effectively and continuously.

3. Interconnect the device with other systems through the implementation of industrial protocols, for information gathering and to facilitate the creation of automation networks.

- **Compliance**: The I2C industrial communication protocol has been integrated for communication with sensors and peripherals, and the MQTT protocol for communication with data management systems, in this case, Node-RED. This integration facilitates automation networking and information gathering.

4.2.2 Specific Objectives

1. Design the PLC electronic circuit, capable of operating in various industrial environments.

   - **Compliance**: was designed and built a robust electronic circuit suitable for use in an industrial environment. Its working place is inside an electrical panel where it coexists with 380V power lines, contactors, motor protectors and other electrical components that generate important electromagnetic fields. Under these conditions, no type of interference or error has been detected in the operation of the system, which demonstrates the degree of robustness achieved. No specific tests have been considered with respect to vibrations, electromagnetic interference, temperature, humidity and environmental dust.

2. Select components and design the printed circuit board (PCB) for the controller manufacturing, considering cost and efficiency aspects.

   - **Compliance**: Both component selection and PCB design have been carried out with a focus on efficiency and low cost, resulting in a system with good performance at a very low production cost.

3. Implement digital and analog input modules, as well as relay and optocoupled output modules, to cover various control needs.

   - **Compliance**: Digital inputs and relay outputs were implemented, all of them opto-coupled to provide robustness. Analog inputs were not considered since the system's sensor readings are digital. On the other hand, opto-coupled outputs were not developed because the switching speed and operating voltage requirements of the actuators are efficiently covered by the relay output module.

4. 3D modeling of the PLC housing, ensuring ergonomic and functional design

- **Fulfillment:** The PLC housing was modeled and then manufactured through 3D printing, achieving an ergonomic and functional design that protects the electronic components and facilitates its installation and handling in industrial environments.

5. Integrate I2C and UART communication protocols to enable efficient interconnection with other devices and systems.

   - **Compliance:** The I2C protocol was successfully integrated in the communication between the PLC and the LCD, and between the PLC and the industrial communication module. The UART protocol, which is intended to communicate two devices, was not considered as it requires the addition of multiplexers to be able to add more devices in the network.

6. Use a microcontroller from the range of options offered by Arduino and similar, optimizing performance at low cost and taking advantage of its wide acceptance in the community.

   - **Fulfillment:** The selection of the Arduino Nano microcontroller as the core of the PLC has offered good performance at a very affordable price. On the other hand, the wide acceptance and diffusion of Arduino in the community facilitated the development and implementation of the PLC with this technology.

7. Implement an industrial communication module to transmit data from the sensors and the PLC to a server, which will allow remote monitoring of the process from any device with an internet connection.

   - **Fulfillment**: A prototype of an industrial communication module was developed that uses the Ethernet and MQTT protocol to connect to NodeRED, where process variable data is managed in real time. This implementation has allowed remote monitoring from any device with internet connectivity.

4.3 Project impact and relevance

4.3.1 Innovation

The development of this project has provided a new automation solution for our region, representing a significant advance in the democratization of industrial control. Its easy programming, community support, free development thanks to free software, and the abundance of resources for the expansion of the system encompass a solution with a different work philosophy to that found in the market.

4.3.2 Cost Reduction

The low cost of the system that has been developed allows companies to reduce the initial investment required to automate their processes, which facilitates the advancement of technology in sectors with limited resources. This cost reduction is reflected not only in the initial purchase of the device, but also in maintenance and upgrade costs, since open technology components and resources are widely available. In addition, the cost associated with the use of software services is completely eliminated.

4.3.3 Flexibility and Customization

The use of Arduino as the core PLC platform provides great design flexibility to incorporate different features and adapt the system to different automation needs, either with hardware modifications, software modifications or by incorporating additional modules.

4.3.4 Industrial and Domestic Applications

While the developed PLC has proven to be able to operate in a real industrial environment, it is impossible to guarantee stable behavior under more aggressive environmental conditions. On the other hand, the communication module shows great potential for data management and device interconnectivity.

These considerations make the PLC, in its current state, suitable for working in less demanding environments, such as home automation. In the industrial environment, thanks to the interconnectivity of the device, automation networks can be developed in conjunction with proprietary PLCs, where the open source PLC takes care of non-critical tasks such as information gathering and data management.

4.3.5 Interconnectivity and Scalability

The integration of I2C and MQTT industrial communication protocols, along with Node-RED for IoT device management, makes automation networking possible, adding the ability to collect and analyze data for remote process monitoring and control.

4.4 Limitations and areas for improvement

1. Incorporation of analog inputs

The PLC developed does not have analog inputs because they have not been required for the automation of the reverse osmosis equipment, but it represents a future limitation for the incorporation of new analog sensors. Therefore, it is important to consider the implementation of 4-20mA and 0-10V analog readings, which are the most used in the industrial field.

2. Handcrafted PCB fabrication

The PCB has been fabricated using the thermal transfer method, which is a time-consuming and manually intensive process, resulting in variable and erratic part finishes. A potential improvement is to consider having the PCB fabricated by a specialized supplier, taking into account the quality of the product and the impact on the fixed costs of the device.

3. Assembly of electronic components

Similar to the handmade PCB manufacturing, assembling the electronic board by hand is a labor intensive task, requiring precision and dedication to position and solder each electronic component in place. This process is not scalable for large-scale production, and introduces variability in product quality, so the manufacture of the entire electronic board by an external supplier should be considered.

4. Cost of casing manufacturing

From the Figure N° 87Hardware costs - communication moduleWe can identify that the largest percentage of the PLC hardware cost is destined to the manufacture of the housing. 3D printing gives us great design flexibility and allows us a small-scale production, but it is not the most cost-effective option if the goal is a larger production, in which case we should investigate the manufacture of parts by plastic injection or other methods. A more immediate solution is to redesign the PLC format, seeking to reduce the housing area and decrease the wall thickness without compromising the strength of the parts.

5. Cybersecurity

With the industrial communication module, the PLC has become part of the internet network, thus being exposed to cyber attacks. Therefore, to ensure the integrity and reliability of the control and monitoring system, the relevant security measures must be comprehensively addressed.

6. Tests and trials

The PLC has been put into operation and has demonstrated proper performance within its working environment, but has not been subjected to specific tests or trials to evaluate its performance under extreme conditions. It is of great importance to consider testing the device against vibration, electromagnetic interference, temperature, humidity and dust from the environment.

7. Watchdog logic

A watchdog system has not been considered in the developed PLC. It is of utmost importance that in future developments a watchdog logic is implemented to supervise the PLC operation, restarting it automatically or bringing it to a safe state in case of failures or critical errors, to guarantee the continuity, stability and safety of the industrial operations.

4.5 Personal reflection

The development of this project has been an enriching experience that has expanded my knowledge in various areas of mechatronics engineering. On the one hand, disciplines such as electronics, computer-aided design and programming were deepened, and specific topics such as automation, communication protocols and the Internet of Things were addressed. Strong theoretical analysis followed by implementation has strengthened the learning. The essence of mechatronics engineering has been revealed in the process of integrating different disciplines in pursuit of finding an effective solution for a particular need.

4.6 Final conclusion

The project arose in response to a specific automation need, which in the course of development revealed a general need for accessible and open technology. The final result of the device demonstrates that such a solution is possible, advancing the democratization of industrial automation and bringing the technology to new areas and applications.

It has been a great progress in automation and engineering, made all the more interesting by the knowledge that there is still much to advance and evolve.

Glossary

AES	Advanced encryption standard for wireless local area networks established in 802.11i that offers a higher level of security than that found in the current WPA security standard.
ADC	Analog-to-digital converter, a device that converts analog signals into digital data.
Cybersecurity	A set of measures and practices designed to protect systems and networks from cyber attacks.
Industrial Communication	Methods and protocols used to enable data transmission between devices and systems in an industrial environment.
Dashboard	Graphical interface that presents information and data in a visual way, facilitating monitoring and analysis.
IoT	(Internet of Things). Network of interconnected devices that can communicate with each other and with other systems through the Internet.
MQTT	(Message Queuing Telemetry Transport). Lightweight messaging protocol ideal for IoT communication. It allows the transmission of data between devices efficiently and securely, facilitating the monitoring and remote control of systems.
Node-RED	Development tool that facilitates the integration of hardware, APIs and online services through a visual flow editor.
PCB	(Printed Circuit Board). Printed circuit board used to assemble and connect electronic components.
PID	(Proportional-Integral-Derivative). Controller used in control systems to regulate process variables. It adjusts the system output based on the current error, the integral of the past error and the derivative of the future error.
Silk-screen printing	Printing process used in the manufacture of PCBs to apply ink layers on the boards.
SM	Sensor module, a component that includes specific sensors to measure physical variables and convert them into electrical signals for processing. Analog SMs contain ADCs for signal conversion.
Thermotransfer	A method of PCB etching that uses heat to transfer an ink design to a blank board.
WPA	(Wi-Fi Protected Access). Security standard for wireless networks.

Bibliographic References

[1] W. Bolton, MECATRONICS: electronic control systems in mechanical and electrical engineering, D.F., Mexico: Alfaomega, 2013.

[2] IEEC UNED, "MASTER DEGREE: Industrial Systems Engineering," [Online]. Available: http://www.ieec.uned.es/investigacion/dipseil/pac/archivos/informacion_de_referencia_ise6_1_1.pdf. [Last access: 13 March 2024].

[3] R. Zurawski, Industrial Communication Technology Handbook, San Francisco, California, USA: CRC Press, 2015.

[4] U. N. D. E. E. A. DISTANCIA, Comunicaciones Industriales - Principios Basicos, Madrid: Librería UNED, 2007.

[5] UNIVERSIDAD NACIONAL DE EDUCACIÓN A DISTANCIA, Industrial Communications: Basic Principles, Madrid: UNED, 2007.

[6] A. R. Penin, Industrial Communications, Barcelona: Marcombo S.A., 2008.

[7] UNIVERSIDAD NACIONAL DE EDUCACION A DISTANCIA, Comunicaciones Industriales: Sistemas distribuidos y aplicaciones, Madrid: UNED, 2007.

[8] J. R. Alejandro Amaya, PROTOCOLS AND TRANSMISSION STANDARDS USED BY INDUSTRIAL INSTRUMENTS, INACAP, 2016.

[9] WESCO-ANIXTER, "Comparing wireless communication protocols," [Online]. Available: https://www.anixter.com/es_mx/resources/literature/techbriefs/comparing-wireless-communication-protocols.html?timeout=true.

[10 Amazon AWS, "What is MQTT," [Online]. Available: https://aws.amazon.com/es/what-is/mqtt/. [Last Accessed: 26 June 2024].

[11 FICA-UNSLL, "Curso de Arduino: Módulo protocolos de comunicaciones," Villa Mercedes, 2024.

[12 Red Hat, "What is open source," 24 Jan. 2023. [Online]. Available: https://www.redhat.com/es/topics/open-source/what-is-open-source. [Last access: 15 March 2024].

[13 Open source initiative, "The Open Source Definition," 16 February 2024. [Online]. Available: https://opensource.org/osd. [Last access: 15 March 2024].

[14 Open Source Hardware Association, "Statement of Principles 1.0," 2024. [Online]. Available: https://www.oshwa.org/definition/spanish/. [Last accessed: 15 March 2024].

[15 Arduino, "What is Arduino," [Online]. Available: https://arduino.cl/que-es-arduino/. [Last accessed: 15 March 2024].

[16 Texas Instruments, "LM2596," [Online]. Available: https://www.ti.com/product/es-mx/LM2596. [Last accessed: 18 March 2024].

[17 Unit Electronics, "PC817 DIP-4 Optocoupler," [Online]. Available: https://uelectronics.com/producto/optoacoplador-pc817-dip-4/. [Last accessed: 18 March 2024].

[18 Arduino, "Arduino Professional," 2024. [Online]. Available: https://store-usa.arduino.cc/pages/professional. [Last accessed: 20 March 2024].

[19 Finder, "FINDER OPTA: THE FIRST PROGRAMMABLE LOGIC RELAY," 01 December 2022. [Online]. Available:

https://www.findernet.com/es/uruguay/news/finder-opta-el-primer-rele-logico-programable/. [Last accessed: 20 March 2024].

[20 Arduino Store, "Introducing Arduino Opta," [Online]. Available: https://store-usa.arduino.cc/pages/opta. [Last accessed: 20 March 2024].

[21 Industrial Shields, "Industrial Controllers and Panel PCs based on Open Source Hardware," [Online]. Available: https://www.industrialshields.com/es_ES. [Last accessed: 20 March 2024].

[22 Controllino, "INDUSTRIAL AUTOMATION MEETS OPEN SOURCE," [Online]. Available: https://www.controllino.com/. [Last accessed: 21 March 2024].

[23 Electroall, "Electroallweb," [Online]. Available: https://www.electroallweb.com/. [Last accessed: 21 March 2024].

[24 El profe Zurco, "Blog El profe Zurco," [Online]. Available: https://elprofezurco.blogspot.com/2023/11/mini-plc-arduino-atmega128a-au-v10.html. [Last accessed: 21 March 2024].

[25 Arduino.cc, "Arduino Software," [Online]. Available: https://www.arduino.cc/en/software. [Last accessed: 21 March 2024].

[26 Codesys, "Codesys Store," [Online]. Available: https://store.codesys.com/en/. [Last accessed: 2022 March 2024].

[27 Autonomy, "OpenPLC," [Online]. Available: https://autonomylogic.com/. [Last accessed: 22 March 2024].

[28 EasyEDA, "Easy-to-use & Free PCB Design Software," [Online]. Available: https://easyeda.com/. [Last accessed: 01 April 2024].

[29 EMQX, "Free Public MQTT Broker," [Online]. Available: https://www.emqx.com/en/mqtt/public-mqtt5-broker. [Last accessed: 2024 July 04].

[30 J. SALAZAR, WIRELESS NETWORKS, Czech Republic: TechPedia .

Annex(es)

1. PLC electronic schematic

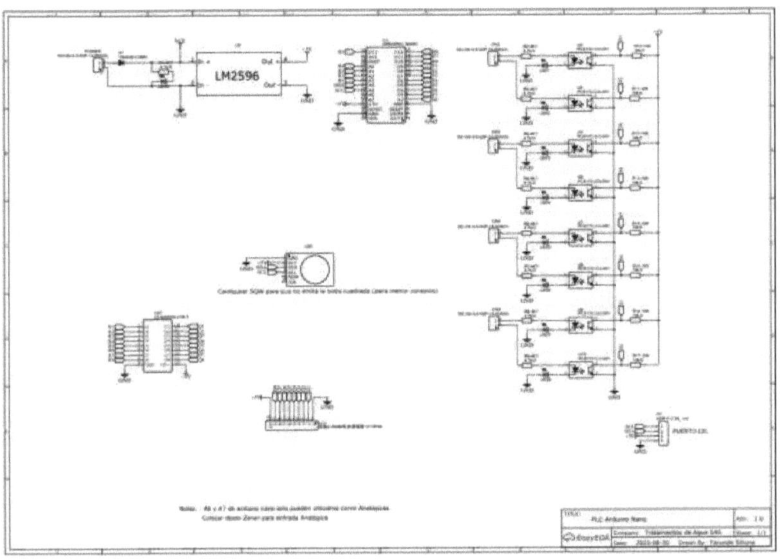

2. PCB Layout - Top screen printing layer

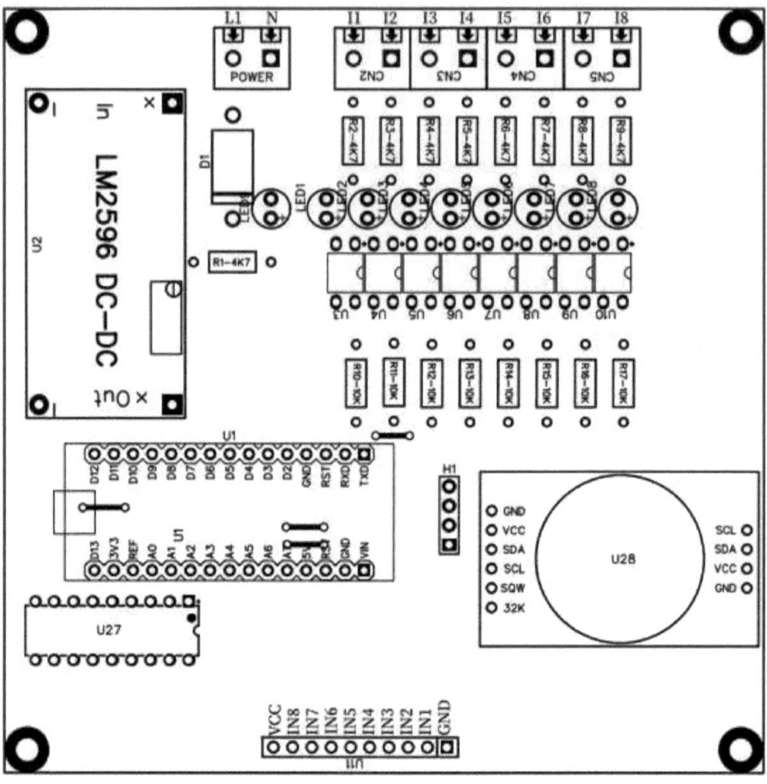

3. PCB Layout - Bottom Layer

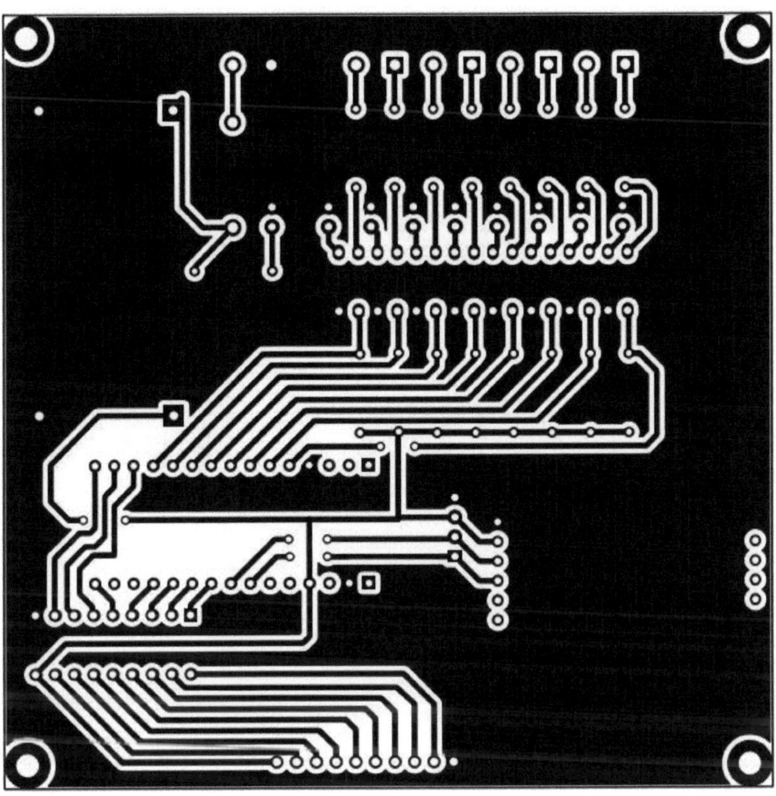

4. PLC assembly drawing

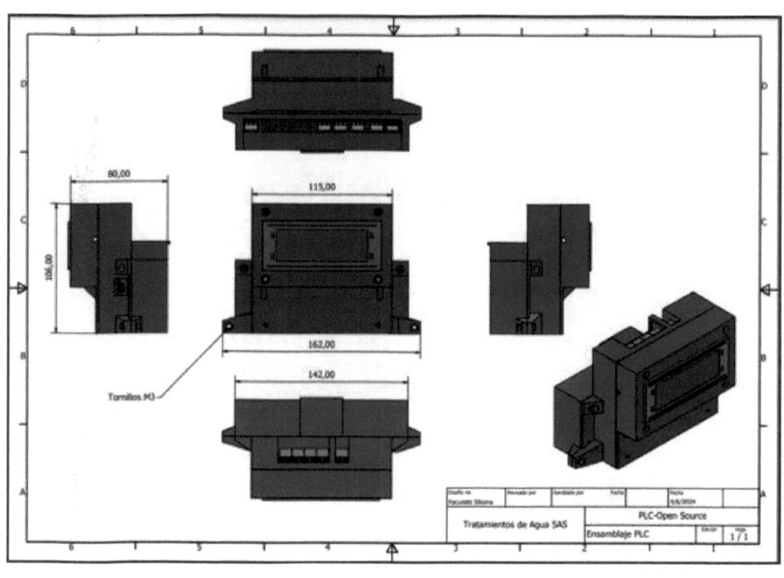

5. Electronic schematic of industrial communication module

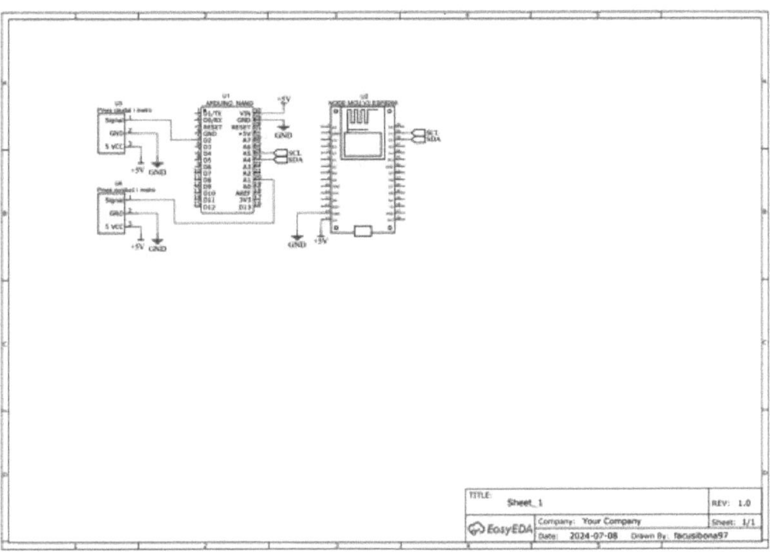

Printed by Books on Demand GmbH, Norderstedt / Germany